Journal of Applied Logics - IfCoLog Journal of Logics and their Applications

Volume 13, Number 4

August 2026

Disclaimer

ISBN 978-1-84890-523-8
ISSN (E) 2631-9829
ISSN (P) 2631-9810

College Publications
Scientific Director: Dov Gabbay
Managing Director: Jane Spurr

http://www.collegepublications.co.uk

EDITORIAL BOARD

Scope and Submissions

This journal considers submission in all areas of pure and applied logic, including:

pure logical systems
proof theory
constructive logic
categorical logic
modal and temporal logic
model theory
recursion theory
type theory
nominal theory
nonclassical logics
nonmonotonic logic
numerical and uncertainty reasoning
logic and AI
foundations of logic programming
belief change/revision
systems of knowledge and belief
logics and semantics of programming
specification and verification
agent theory
databases

dynamic logic
quantum logic
algebraic logic
logic and cognition
probabilistic logic
logic and networks
neuro-logical systems
complexity
argumentation theory
logic and computation
logic and language
logic engineering
knowledge-based systems
automated reasoning
knowledge representation
logic in hardware and VLSI
natural language
concurrent computation
planning

This journal will also consider papers on the application of logic in other subject areas: philosophy, cognitive science, physics etc. provided they have some formal content.

Submissions should be sent to Jane Spurr (jane@janespurr.net) as a pdf file, preferably compiled in LaTeX using the IFCoLog class file.

Contents

Reconfigurable Transition Systems – ReacTS 2024: Editorial Article

Raul Fervari
Consejo Nacional de Investigaciones Científicas y Técnicas (CONICET) and Universidad Nacional de Córdoba, FAMAF, Argentina

Manuel A. Martins
CIDMA, University of Aveiro, Portugal

Abstract

This article introduces the special issue collecting extended versions of papers presented at the 1st International Workshop on Reconfigurable Transition Systems (ReacTS 2024), held in Aveiro, Portugal, on November 4–5, 2024. The workshop aims to highlight the importance of reconfigurable and reactive systems as an interdisciplinary research area, bringing together computer scientists, mathematicians, philosophers, and researchers from related fields. This interdisciplinary nature is reflected in the diversity of contributions accepted both to the workshop and to this special issue. We hope that ReacTS 2024 marks the beginning of a regular series of meetings and fosters many further exciting interactions and collaborations.

Foreword

Reconfigurable Transition Systems (RTSs) are dynamic relational structures (i.e., graphs) that evolve during their execution, in the sense that their accessibility relation, set of nodes, or labelling may change when their edges are traversed. These structures have proven suitable for compactly representing complex reactive and reconfigurable behaviours. Indeed, the ability to react or adapt in response to certain events is a distinctive feature of many systems and phenomena. An autonomous vehicle that changes its route due to a newly announced strike, the behaviour of a software component after memory disposal, or a DNA mutation resulting from a viral infection are examples that illustrate the importance of modelling change in a given situation. Such practical use cases have sparked the interest of the logic community in the study of RTS variants and in the development of formal methods for reasoning about these phenomena.

This special issue collects extended papers of the first 'International Workshop on Reconfigurable Transition Systems: Semantics, Logics and Applications' ReacTS 2024, which took place in November 2024, in Aveiro, Portugal. This workshop aims to bring together the whole community of researchers working on different ways to model reconfigurable and reactive systems from a formal perspective. This includes theoretical approaches (like hybrid logics, reactive frames, model-update logics), or formalisms designed for specific purposes (like separation logic in software verification, dynamic epistemic logic in AI planning, and others). More broadly, the workshop seeks to foster novel approaches and applications, while promoting a shared perspective on the discipline.

This volume contains four accepted papers. All submissions went through an additional reviewing phase to ensure a high quality for this collection. We briefly summarize each contribution below.

"Resolving Distributed Belief", authored by John I. Lindqvist, Fernando R. Velásquez-Quesada and Thomas Ågotnes, proposes three different approaches for udpating beliefs of a group of agents while maintaining their consistency.

In *"Paraconsistent Reactive Graphs: Their Semantics and Conflict Resolution Policies"*, Juliana Cunha, Alexandre Madeira, and Luís Soares Barbosa investigate an extension of Reactive Graphs enriched with paraconsistency on graph edges, specifically designed to address vagueness and inconsistency in dynamic systems.

"Beyond the Spell: A Dynamic Logic Analysis of Misdirection", authored by Benjamin Icard and Raul Fervari, introduces a framework based on Dynamic Epistemic Logic for formalizing deception actions between agents, arising from both verbal and observational sources.

Finally, *"Intra and Inter-Paradigm Reconfigurations with ZX: An Exercise"*, by Bruno Jardim, Jaime Santos and Luís Soares Barbosa, studies how staggered quantum walks can be modelled through a graph reconfiguration process and implemented within the measurement-based paradigm.

We would like to thank all authors who submitted their work, the reviewers for their careful and constructive evaluations, and the journal's editorial team for their support throughout the preparation of this special issue.

July 2026

Raul Fervari
Manuel A. Martins
Guest Editors

Received June 2026

Resolving Distributed Belief

John I. Lindqvist
Department of Information Science and Media Studies, University of Bergen, Norway
John.Lindqvist@uib.no

Fernando R. Velázquez-Quesada
Department of Information Science and Media Studies, University of Bergen, Norway
Fernando.VelazquezQuesada@uib.no

Thomas Ågotnes
Department of Information Science and Media Studies, University of Bergen, Norway
School of Philosophy, Shanxi University, Taiyuan, China
Thomas.Agotnes@uib.no

Abstract

In multi-agent systems, a central informational action is that in which agents communicate to others what they know or believe. In 'standard' epistemic logic, this action has been modelled as a transformation of the underlying relational structure: agents in the communicating group are assigned the relation corresponding to their distributed information in the initial model (other agents' relations remain unchanged). While this idea works well when the passed information is knowledge, problems arise when it concerns beliefs: consistent agents might become inconsistent. This paper introduces and analyzes three belief-update operations designed to address this issue: cautious belief resolution, patched cautious belief resolution, and belief sharing. The first two use maximally consistent subgroups of agents, to ensure that consistency is preserved. The third provides a novel mechanism that captures the 'safe' aspects of belief communication—namely, the sharing and accepting of agents' beliefs about their own beliefs.

1 Introduction

Epistemic logic (*EL*; ,) is a framework for representing and reasoning about what different agents know or believe—not only about the world, but also about their own information and that of other agents. Its 'dynamic' extensions (*DEL*; , ,) focus on representing and reasoning about informational changes/events. As such, *EL* and *DEL* are useful tools for reasoning about knowledge, belief, and information change, all those important topics for areas where multi-agent interaction is of interest (e.g., A.I., Computer Science, Economics, Philosophy).

A particularly important informational action is that of some agents communicating their individual information to some others, and a crucial aspect of this is the way the received information is combined. This general idea has been studied from various perspectives, depending on the information that is being passed and the way it is represented. One example is the case of *preferences* represented by ordering relations, which has led to the study of voting systems (e.g., Part 1 of). Another is the case of beliefs represented syntactically as finite belief bases, which has given rise to belief merging (e.g.,) and judgement aggregation (e.g., ,). Within *EL*, this communication act is closely related to the concept of *distributed knowledge* [, ,], which intuitively describes the logical consequences of the combination of the individual knowledge of the members of a group. When *EL*'s semantics are given by relational 'Kripke' models (structures that represent possible worlds and indistinguishability between them), the relation for the distributed knowledge of a group is given by the *intersection* of individual relations of the group's members. In this way, the group has distributed knowledge of φ at a world w if and only if φ holds in all situations that *no member of the group* can distinguish from w. Most, if not all, *EL*-based studies of this kind of communication rely on this concept.

There are several variations of the described communication action (see, for example, , , , ,). The focus here is the case in which every agent in a group communicates *all* her information with *everybody else* in the group, while all other agents outside the group neither pass nor receive any information. The case in which the passed information is knowledge (i.e., truthful) is studied in Ågotnes and Wáng [ ]. Following the *DEL* strategy, this communication action is represented by a model operation. Its input is the model capturing the agents' initial knowledge, and its output is the model representing their knowledge *after* the action has taken place. In particular, this model operation "resolves" the distributed knowledge of a group: it assigns to every agent in the communicating group the relation describing the group's distributed knowledge, leaving the relation of all other agents as before. Thus, intuitively, what was distributively known among the group of agents becomes the group's members' individual knowledge. This 'simple' strategy works well in the knowledge case: the information that will be combined is truthful; thus it is, as a whole, *consistent*, and it can be 'put together' without any additional consideration. However, when the information to be merged consists of beliefs, two consistent agents might have opposite opinions: one might believe a given φ and the other might believe $\neg\varphi$.

This paper studies three belief-update operations designed to address this issue. Relying on the idea of *maximal consistency*, the first two proposals define an operation that generalises the "resolving distributed knowledge" discussed above. The use of maximal consistency is a well-known approach to handle conflicting information (e.g., 4, 11, 23, 30, 35, 42). Here the idea is to use maximally consistent *subgroups of agents*: the new accessibility relation of an agent is not defined by the intersection relation for the entire group, but rather given in terms of the intersection relation for the maximal consistent subgroups she belongs to. In this way, *cautious belief resolution* preserves consistency. However, other belief properties—namely positive and negative introspection—may still be lost. For that reason, a *patched* cautious belief resolution is also introduced and studied. This patched version can be understood as imposing additional restrictions on the information accepted by the agents, to ensure that their initially 'rational' beliefs remain that way.

While beliefs remain rational after patched resolution, the restrictions imply that agents 'reject' information that, intuitively, should be accepted—namely, information about the beliefs of the other agents involved in the communication process. Indeed, while an agent a might reasonably reject agent b's beliefs about the world when they conflict with hers, all else being equal, rejecting b's claims *about b's own beliefs* seems less justified—even when they conflict with what a's beliefs *about b's beliefs*. To handle this, a third update—*belief sharing*—is introduced. Instead of using the intersection relation to capture the information update, this operation redirects the accessibility relation—determining which situations cannot be distinguished—of the sharing agents to reflect their actual information state throughout the model. Intuitively, this captures the first step of belief resolution: agents discover what the sharing agents believe. After such a sharing update, a group can 'safely' perform cautious belief resolution: once belief sharing has corrected agents' higher-order beliefs about one another, cautious resolution preserves both positive and negative introspection, as well as consistency.

Outline. Section 2 presents the preliminaries, including the action of knowledge communication captured by knowledge resolution, showing also what happens when the truthfulness assumption is dropped. Section 3 begins by introducing *cautious belief resolution*, showing that it preserves belief consistency (Proposition 3.1). The language of distributed belief along with the cautious belief resolution modality is shown to be equally expressive to its static fragment (Theorem 4), and an axiomatization is provided via reduction (Theorem 3). However, this operation does not preserve introspection. Thus, a further refinement—*patched cautious belief resolution*—is introduced. It is shown not only to preserve the properties of belief (Proposition 3.3), but also to ensure that, while an agent might in some cases 'forget' or 'ignore' the information she receives, she will not forget information she previously held (Proposition 3.4). It is shown that a modality for this operation increases the expressivity of the language of distributed belief—both in arbitrary models and in models that are serial, transitive and Euclidean (Corollary 3.1)—and axiomatization is discussed. Section 4 introduces *belief sharing* and provides an initial exploration of its properties, in particular its effects on beliefs in belief models (Proposition 4.5). It is shown that belief sharing, when combined with cautious belief resolution, preserves the belief properties (Proposition 4.6), and a problem that arises because of a mismatch between distributed information and relational models is discussed. Section 5 summarizes the content and outlines directions for future research.

This paper extends the conference paper [ ], and thus it contains some material presented there in preliminary form. In particular, the definitions of cautious belief G-resolution (Definition 3.3) and patched cautious belief G-resolution (Definition 3.4), together with initial results concerning their preservation of belief properties, were first introduced there. The present text substantially extends that. It introduces modal operators capturing the effects of these updates, providing a systematic investigation of the resulting languages, including expressivity and axiomatization. In addition, all material in Section 4 is new.

2 Preliminaries

Throughout this paper, let P be a countable set of atoms and $A \neq \varnothing$ a finite set of agents.

Definition 2.1 (Model). *A model is a tuple $\mathcal{M} = \langle W, \{R_i \mid i \in A\}, V \rangle$ where $W \neq \varnothing$ is a domain (its elements called [possible] worlds), each $R_i \subseteq W \times W$ is a binary relation (for a non-empty $G \subseteq A$, define $R_G := \bigcap_{i \in G} R_i$) and $V : P \to \wp(W)$ is an atomic valuation. A* pointed model *is a pair $(\mathcal{M}, w)$ with $\mathcal{M}$ a model and w a world in it.*

Take a model $\langle W, \{R_i \mid i \in A\}, V \rangle$ and a world $w \in W$. For an agent $i \in A$, the set $R_i(w) := \{v \in W \mid R_i wv\}$ is called i's conjecture set *(alternatively, i's* belief state*); for a non-empty $G \subseteq A$, the set $R_G(w) := \{v \in W \mid R_G wv\} (= \bigcap_{i \in G} R_i(w))$*

is called G's conjecture set. *If a model is such that all relations in* $\{R_i \mid i \in A\}$ *have a certain property (e.g., reflexivity, transitivity, Euclidicity, seriality), the model is said to have the property. In particular, a reflexive, transitive and Euclidean model is called a* knowledge model, *and a serial, transitive and Euclidean model is called a* belief model. ◀

Pointed models can be described by various modal languages. This paper will rely on the following one.

Definition 2.2 (Language $\mathcal{L}_{\Box,\mathrm{D}}$). *The language* $\mathcal{L}_{\Box,\mathrm{D}}$ *extends the propositional one (with* P *as its basic set of atoms, including the constant* $\top$*, and with* $\neg$ *and* $\wedge$ *as its basic operators) with two sets of modalities: those in* $\{\Box_i \mid i \in A\}$ *for describing the information of individuals, and those in* $\{\mathrm{D}_G \mid \varnothing \subset G \subseteq A\}$ *for describing the distributed information of groups. For the semantic interpretation, let* $(\mathcal{M}, w)$ *be a pointed model with* $\mathcal{M} = \langle W, \{R_i \mid i \in A\}, V\rangle$*. The semantics for Boolean constants and Boolean operators is as usual. For the rest,*

$$\begin{array}{lll} \mathcal{M}, w \vDash \Box_i\varphi & \textit{iff} & \mathcal{M}, v \vDash \varphi \textit{ for all } v \in W \textit{ such that } R_iwv, \\ \mathcal{M}, w \vDash \mathrm{D}_G\,\varphi & \textit{iff} & \mathcal{M}, v \vDash \varphi \textit{ for all } v \in W \textit{ such that } R_Gwv. \end{array}$$

It is said that a formula φ *is true at* $(\mathcal{M}, w)$ *iff* $(\mathcal{M}, w) \vDash \varphi$*. It is said that* φ *is valid (notation:* $\vDash \varphi$*) iff it is true in every pointed model (i.e., in every world of every model).* ◀

The notion of consistency, made precise below, plays an important role.

Definition 2.3 (Consistency). *An agent* $i \in A$ *(respectively, a group* $G \subseteq A$*) is* consistent *at a world* w *iff* $R_i(w) \neq \varnothing$ *(resp.,* $R_G(w) \neq \varnothing$*). Otherwise,* i *(resp.,* G*) is inconsistent.* ◀

Thus, consistency is a local property, defined at a *pointed* model. Within $\mathcal{L}_{\Box,\mathrm{D}}$, this notion has a syntactic characterisation: an agent i (resp., a group G) is *consistent* at $(\mathcal{M}, w)$ iff $\mathcal{M}, w \vDash \neg\Box_i\bot$ (resp., $\mathcal{M}, w \vDash \neg\,\mathrm{D}_G\,\bot$).

The information of an individual agent usually comes in two flavours.[1] On the one hand, it can be taken to be *knowledge*. In such cases, the modality $\Box_i$ is replaced by K_i (so $\mathrm{K}_i\,\varphi$ is read as *"agent i knows that φ holds"*). More importantly, one requires *knowledge* models, under which each agent's knowledge satisfies truthfulness ($\vDash \mathrm{K}_i\,\varphi \rightarrow \varphi$, the reason being reflexivity) as well as positive and negative introspection ($\vDash \mathrm{K}_i\,\varphi \rightarrow \mathrm{K}_i\,\mathrm{K}_i\,\varphi$ and $\vDash \neg\,\mathrm{K}_i\,\varphi \rightarrow \mathrm{K}_i \neg\,\mathrm{K}_i\,\varphi$, the reason being, respectively, transitivity and Euclidicity). On the other hand, the information can be taken to

[1]Still, one can find in the literature the discussion of further epistemic/doxastic notions (see, e.g., [38], [30]).

be *beliefs*. In such cases, the modality $\Box_i$ is replaced by B_i (so $\mathrm{B}_i\,\varphi$ is read as *"agent i believes that φ holds"*). More importantly, one requires *belief* models, under which each agent's beliefs satisfy consistency ($\models \neg\,\mathrm{B}_i\,\bot$, the reason being seriality) as well as positive and negative introspection ($\models \mathrm{B}_i\,\varphi \rightarrow \mathrm{B}_i\,\mathrm{B}_i\,\varphi$ and $\models \neg\,\mathrm{B}_i\,\varphi \rightarrow \mathrm{B}_i\,\neg\,\mathrm{B}_i\,\varphi$).

The information of groups of agents comes in a larger variety of flavours. First, one can distinguish between the group's knowledge and its beliefs. But even if one of them is fixed (by fixing the information of the group's members), there are still several possibilities. One might say that a group knows/believes a given φ when *everybody* in the group does so, but also when *someone* in the group does so (see for the latter). One might even talk about a group of agents having *common* knowledge/belief on φ [ ]: everybody in the group knows/believes φ, everybody in the group knows/believes that everybody in the group knows/believes φ, and so on.

The group notion this paper relies on is that of *distributed* knowledge/beliefs [, ,]. Intuitively, a group has distributed knowledge/beliefs of φ if and only if φ is a logical consequence of the combination of the individual knowledge/beliefs of the group's members. Within the *EL* setting, where uncertainty is used to define knowledge/beliefs, this intuition has a natural representation: at a world w a group G has distributed knowledge/beliefs of φ if and only if φ holds in all the situations *no one in the group* can distinguish (i.e., in all the situations that *all the members of the group* can reach) from w. In other words, a group G has distributed knowledge/beliefs of φ at a pointed model $(\mathcal{M}, w)$ iff $\mathcal{M}, v \models \varphi$ for all $v \in R_G(w)$, and thus iff $\mathcal{M}, w \models \mathrm{D}_G\,\varphi$.

Since the relation R_G is defined as the intersection of the relations $\{R_i \mid i \in G\}$, the properties of a group's distributed information depend on the properties of the information of the group's members. When the information of all agents in G is knowledge, $\mathrm{D}_G\,\varphi$ satisfies truthfulness as well as both positive and negative introspection (intersection preserves reflexivity, transitivity and Euclidicity); when the information of all agents in G is beliefs, $\mathrm{D}_G\,\varphi$ satisfies positive and negative introspection, but it might not satisfy consistency (intersection does not preserve seriality).[2]

Finally, it should be noted that this *EL* definition of distributed knowledge/-beliefs is not without faults. A crucial aspect is that it does not fully align with the underlying intuition: because the models permit different possible worlds to satisfy exactly the same formulas, there are pointed models in which $\mathrm{D}_G\,\varphi$ holds even though φ is not a logical consequence of the union of all formulas known by the members of G [, ,]. There are different strategies for dealing with this, including restricting the class of models [, ,] and also providing alternative

[2]See for a systematic study of the properties of different forms of group knowledge/beliefs.

semantic definitions [16, 18, 37]. Still, the *EL* definition also has useful features: among others, its associated relation R_G is intuitively appealing, it turns D_G into a normal modality (which distributes over implications and satisfies the generalisation rule) that inherits the properties of knowledge from its members, and it allows natural representations of actions for knowledge communication (see Subsection 2.1 below).

Expressivity Note that the modalities for the information of individual agents are not needed. Indeed, the individual information of an agent is the same as the distributed information of the group that has the agent as its lone member ($\models \Box_i\varphi \leftrightarrow \mathrm{D}_{\{i\}}\,\varphi$). Thus, the languages $\mathcal{L}_{\Box,\mathrm{D}}$ and $\mathcal{L}_{\mathrm{D}}$ are equally expressive.

Still, in some cases, comparing the expressivity of two languages is not straightforward. In such situations, it is useful to have a semantic notion that connects two pointed models when they cannot be distinguished by a language's formulas. For the basic modal language $\mathcal{L}_{\Box}$, the notion of *bisimulation* plays this role (see, e.g., 9, Definition 2.18 and Theorem 2.20). When the modality for distributed knowledge is included, one needs rather the notion of *collective* bisimulation [36], which expands on a standard bisimulation by asking for the **forth** and **back** clauses to be satisfied not only by all relations for individuals (those in $\{R_i \mid i \in A\}$) but also by all relations for groups of agents (those in $\{R_G \mid \varnothing \subset G \subseteq A\}$).

Definition 2.4 (Collective bisimulation). *Let $\mathcal{M} = \langle W, \{R_i \mid i \in A\}, V\rangle$ and $\mathcal{M}' = \langle W', \{R'_i \mid i \in A\}, V'\rangle$ be two models. A non-empty relation $Z \subseteq W \times W'$ is a* collective bisimulation *between $\mathcal{M}$ and $\mathcal{M}'$ if and only if every $(u, u') \in Z$ satisfies the following.*

- ***Atoms**. For every $p \in P$: $u \in V(p)$ if and only if $u' \in V'(p)$.*
- ***Forth**. For every $\varnothing \subset G \subseteq A$ and every $v \in W$: if $R_G uv$ then there is $v' \in W'$ such that both $R'_G u'v'$ and $(v, v') \in Z$.*
- ***Back**. For every $\varnothing \subset G \subseteq A$ and every $v' \in W'$: if $R'_G u'v'$ then there is $v \in W$ such that both $R_G uv$ and $(v, v') \in Z$.*

Write $\mathcal{M} \leftrightarrows_{\mathrm{D}} \mathcal{M}'$ if and only if there is a collective bisimulation between $\mathcal{M}$ and $\mathcal{M}'$. Write $(\mathcal{M}, w) \leftrightarrows_{\mathrm{D}} (\mathcal{M}', w')$ if and only if some witness for $\mathcal{M} \leftrightarrows_{\mathrm{D}} \mathcal{M}'$ contains the pair (w, w'). ◀

Below we introduce notation for equivalence w.r.t. formulas of a language.

Definition 2.5 ($\mathcal{L}$-equivalence). *Let $\mathcal{L}$ be a language whose formulas can be evaluated in pointed models. Two pointed models $(\mathcal{M}, w)$ and $(\mathcal{M}', w')$ are* $\mathcal{L}$-equivalent

PR: $\vdash \varphi$ for φ a propositionally valid scheme	
MP: If $\vdash \varphi$ and $\vdash \varphi \to \psi$ then $\vdash \psi$	
K_{D}: $\vdash \mathrm{D}_G(\varphi \to \psi) \to (\mathrm{D}_G\,\varphi \to \mathrm{D}_G\,\psi)$	G_{D}: If $\vdash \varphi$ then $\vdash \mathrm{D}_G\,\varphi$
M_{D}: $\vdash \mathrm{D}_G\,\varphi \to \mathrm{D}_{G'}\,\varphi$ for $G \subseteq G'$	

Table 1: Axiom system for $\mathcal{L}_{\mathrm{D}}$

(notation: $(\mathcal{M}, w) \leftrightsquigarrow_{\mathcal{L}} (\mathcal{M}', w'))$ *if and only if, for every* $\varphi \in \mathcal{L}$,

$$(\mathcal{M}, w) \models \varphi \quad \textit{if and only if} \quad (\mathcal{M}', w') \models \varphi. \qquad \blacktriangleleft$$

The following proposition shows that collective bisimilarity is indeed useful for our purposes: $\mathcal{L}_{\mathrm{D}}$ is invariant under collective bisimilarity.

Theorem 1 (Collective bisimilarity implies $\mathcal{L}_{\mathrm{D}}$-equivalence)**.** *Let* $(\mathcal{M},w)$ *and* $(\mathcal{M}',w')$ *be two pointed models. If* $(\mathcal{M}, w) \rightleftarrows_{\mathrm{D}} (\mathcal{M}', w')$*, then* $(\mathcal{M}, w) \leftrightsquigarrow_{\mathcal{L}_{\mathrm{D}}} (\mathcal{M}', w')$.

Proof. Proofs showing that a form of structural equivalence implies invariance for a language typically use structural induction on the language's formulas. For the case of collective bisimilarity and $\mathcal{L}_{\mathrm{D}}$ see, e.g., Roelofsen [illegible]. □

Axiomatisation The axiom system in Table 1 (from, e.g., [illegible], [illegible]) characterises the formulas of $\mathcal{L}_{\mathrm{D}}$ valid in pointed models. The behaviour of Boolean operators is characterised by PR and MP. For each modality D_G, while rule G_{D} indicates that it 'contains' all validities, axiom K_{D} indicates that it is closed under modus ponens. (By using the definition $\Box_i\varphi := \mathrm{D}_{\{i\}}\,\varphi$, analogous rule and axiom can be derived for $\Box_i$.) Additionally, Axiom M_{D} states that the operator is monotone on the group of agents: if φ is in G's distributed information, then it is also in the distributed information of any larger group.

Theorem 2. *The axiom system in Table 1 is sound and strongly complete for* $\mathcal{L}_{\mathrm{D}}$ *w.r.t. the class of all pointed models.* ◀

2.1 Communicating knowledge: the 'natural' approach works

Individual information can change through different actions. Within *DEL*, the effect of each such action is represented by taking the (pointed) model depicting the agents' initial information, then returning the one depicting their information after the action has taken place.

Since the individual information of the agents is potentially different, a natural action to study is one through which some agents communicate their individual information to some others. There are variations of this, depending on who communicates, what is being passed, who receives and what is learnt (if anything) by agents not receiving (examples include 1, 6, 7, 17, 47). One of the simplest (and still interesting) scenarios is that in which all members of a group G convey all their information within themselves, with all other agents passing and receiving nothing (yet noticing that communication within G happened). The case in which the passed information is knowledge corresponds to a *knowledge resolution* [1]. As it is usual in 'knowledge' scenarios, knowledge models are assumed and modalities K_i are used.

Definition 2.6 (Knowledge G-resolution). *Let $\mathcal{M} = \langle W, \{R_i \mid i \in A\}, V\rangle$ be a model; take $\varnothing \subset G \subseteq A$. The* knowledge G-resolved update *of $\mathcal{M}$ is the model $\mathcal{M}[G] = \langle W, \{R_i[G] \mid i \in A\}, V\rangle$ (so, domain and valuation remain the same) in which each $R_i[G]$ is given by*

$$R_i[G] := \begin{cases} R_G(= \bigcap_{b\in G} R_b) & i \in G; \\ R_i & \textit{otherwise.} \end{cases}$$

For describing this operation's effects, add an operator $[G!]$ for each non-empty $G \subseteq A$, with its semantic interpretation given by

$$\mathcal{M}, w \vDash [G!]\varphi \quad \textit{iff} \quad \mathcal{M}[G], w \vDash \varphi$$ ◀

Thus, while the operation leaves the indistinguishability relation for agents not in G untouched (they do not receive what agents in G pass), it assigns to every agent in G the group's distributed information relation (i.e., the intersection of the agents in G's initial relations, as only they communicate).[3] Here are some features of this definition that are worth noting.

- $R_i[G] \subseteq R_i$ for every agent $i \in A$: the action might reduce some agents' uncertainty, but it will never increase it.
- Knowledge models (and thus the properties of knowledge) are preserved: if every R_i is an equivalence relation, so are all relations $R_i[G]$.

[3]Still, because of the nature of the (relational) models, under this definition the information of agents not in G changes too: they get to know that agents in G have passed information within themselves. *Secret* forms of communication (some agents exchange information without the others noticing it) can be also represented: see Baltag and Smets [ ] for this in the context of agents 'taking' rather than 'passing', and Baltag et al. [ ] for this in the context of announcements/observations (i.e., agents receiving information from an *external* source).

- The validities below [ ] make precise what the individual knowledge of the agents is after this form of communication:

$$\text{for } i \in G\text{:} \;\; \models [G!]\, \mathrm{K}_i\, \varphi \leftrightarrow \mathrm{D}_G[G!]\varphi; \qquad \text{for } i \notin G\text{:} \;\; \models [G!]\, \mathrm{K}_i\, \varphi \leftrightarrow \mathrm{K}_i[G!]\varphi. \quad (1)$$

  Thus, agents in G get to know φ after the action iff, before the action, the group G knew, distributively, that the action would make φ true. Agents not in G need a stronger requirement to know φ after the action: they need to know before, on their own, that the action would make φ true.

2.2 Communicating beliefs: the 'natural' approach has issues

Suppose now that the agents' information is not knowledge but rather beliefs; accordingly, one works with belief (serial, transitive and Euclidean) models. How can one define an operation representing the action of belief communication? A first approach is to use the same idea as in the 'knowledge' case (Definition 2.6 above). However, an immediate issue arises: even though transitivity and Euclidicity are preserved by the operation (i.e., by intersections), seriality is not. This means that the operation might yield inconsistent agents, even when none were inconsistent before. More to the point: when consistent agents resolve their beliefs with those of other consistent agents who disagree, they will turn inconsistent.

Example 2.1. *Consider the belief model $\mathcal{M}_1$ (below left). Its $\{a, b\}$-resolved update, the model $\mathcal{M}_1[\{a, b\}]$ (below right), is not serial: at w neither a nor b has an accessible world.*

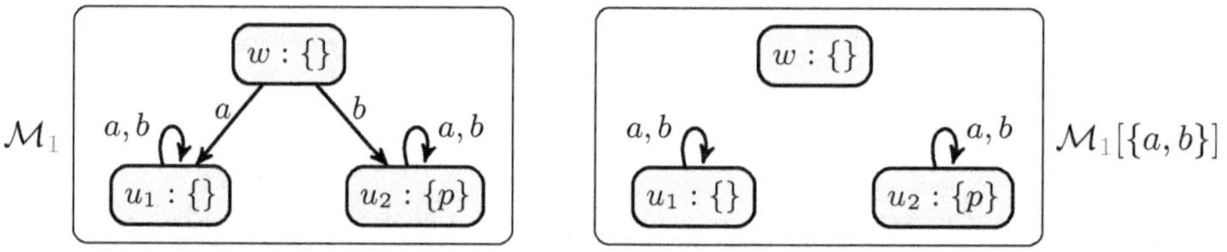

Thus, resolving makes both a and b inconsistent:

$$\mathcal{M}_1, w \models \neg\Box_a\bot \land \neg\Box_b\bot \land [G!](\Box_a\bot \land \Box_b\bot). \qquad ◀$$

Thus, in this belief communication case, a different approach is needed to ensure that the resulting agents remain consistent. Since the issue of preserving consistency has arisen in the literature before, it is useful to look at the different paths taken. A natural idea is to use a form of reliability raking over the communicating agents, to make decisions when conflicts arise (see, e.g., , ,). Still, this additional information might not always be available, and thus lacks generality. Other alternatives arise from the related scenario in which agents receive a *public announcement* [19, 34] contradicting their beliefs. In such cases, while some proposals have advocated for ignoring the information when it is conflicting (e.g., 40), some others make the action impossible unless the resulting model has the required properties [5]. Still, as inspiring as these ideas might be, here the situation is different: the new information comes not from a non-modelled external source, but rather from agents within the system. Thus, an alternative approach is needed.

3 Using maximal consistency

In the knowledge case, the fact that intersections preserve the required relational properties is crucial for both distributed information as well as the G-resolution update. For the first, it makes this distributed information (given as normal modality working on the relation R_G) a 'proper' form of knowledge; for the latter, it guarantees that the properties of individual knowledge are preserved. However, intersections do not preserve seriality, a property typically required for a model representing beliefs. Thus, the distributed belief of a group of consistent agents might end up not being consistent [2], and a resolution update might turn previously consistent agents into inconsistent ones (Example 2.1).

To avoid ascribing inconsistent beliefs to a group of consistent agents, Lindqvist et al. [30] proposes two forms of distributed beliefs, related to the *sceptical* and *credulous* consequence in default logic [35]. Both of them rely on the idea of *maximally consistent* subsets from a given group of consistent agents $G \subseteq A$.

Definition 3.1 (Maximal consistency). *A group $H \subseteq G \subseteq A$ is* maximally consistent w.r.t. *G at a world w (notation: $H \subseteq_w^{max} G$) iff it is consistent at w and, additionally, adding any agent in $G \setminus H$ makes the resulting group inconsistent.* ◀

Thus, at a given world, there might be several groups of agents that are maximally consistent w.r.t. G. Relying on this, Lindqvist et al. [30] defined two forms of distributed beliefs. On a given w, while a group G has *cautious* distributed belief on φ (in the language: $\mathrm{D}_G^{\forall} \varphi$) iff *every* maximally consistent subset of G has distributed belief on φ, the group has *bold* distributed belief on φ (in the language: $\mathrm{D}_G^{\exists} \varphi$) iff *some* maximally consistent subset of G has distributed belief on φ.

Definition 3.2 (Cautious & bold distributed belief). *Let $(\mathcal{M}, w)$ be a pointed model; let $\varnothing \subset G \subseteq A$ be a group. Then,*

$$\mathcal{M}, w \models \mathrm{D}_G^{\forall} \varphi \quad \textit{iff} \quad \mathcal{M}, w \models \mathrm{D}_H \varphi \textit{ for all } H \subseteq_w^{max} G,$$
$$\mathcal{M}, w \models \mathrm{D}_G^{\exists} \varphi \quad \textit{iff} \quad \mathcal{M}, w \models \mathrm{D}_H \varphi \textit{ for some } H \subseteq_w^{max} G.$$
◀

It should be noted that both cautious and bold distributed beliefs are definable within $\mathcal{L}_{\mathrm{D}}$ [20]. For this, note that the abbreviation

$$max_G(H) := \neg \, \mathrm{D}_H \bot \;\wedge \bigwedge_{H \subset H' \subseteq G} \mathrm{D}_{H'} \bot,$$

states that group H is maximally consistent relative to a group G (in words: for all $(\mathcal{M}, w)$, we have $\mathcal{M}, w \vDash max_G(H)$ iff $H \subseteq_w^{max} G$). With it, we have

$$\vDash \mathrm{D}_G^{\forall} \varphi \leftrightarrow \bigwedge_{\varnothing \neq H \subseteq G} \Big(max_G(H) \rightarrow \mathrm{D}_H \varphi \Big) \quad \text{and} \quad \vDash \mathrm{D}_G^{\exists} \varphi \leftrightarrow \bigvee_{\varnothing \neq H \subseteq G} (\neg \, \mathrm{D}_H \bot \wedge \mathrm{D}_H \varphi)$$

As discussed in the referred paper, cautious and bold beliefs deal better with potential inconsistencies. This suggests that the ideas behind them can be used for defining an action through which the agents in G communicate their beliefs among themselves. In the rest of this section, the cautious approach is studied. (A study of the bold approach is left for future work.)

3.1 Cautious belief resolution

Take a group G and an agent $i \in G$. Following the idea of cautious distributed belief, a *cautious* version of the G-resolution action will, intuitively, ascribe a belief to i iff this belief was held distributively by *every* maximally consistent subset of G *she belonged to*. To achieve this in the *EL* setting, the new indistinguishability relation for i at each world w should be defined as reaching those worlds that are reachable by *at least one* maximally consistent subset of G she belongs to (i.e., by the *union* of the worlds reachable by each such subset of G). This ensures that all worlds compatible with the information of *some* relevant subgroup of G are retained in i's conjecture set. As a consequence, a formula will be true in i's updated conjecture set only if it is true at all worlds in the conjecture sets of *all* such subgroups.[4]

Definition 3.3 (Cautious belief G-resolution). *Let* $\mathcal{M} = \langle W, \{R_i \mid i \in A\}, V \rangle$ *be a model; take* $\varnothing \subset G \subseteq A$. *The* cautious G-resolved update *of* $\mathcal{M}$ *is the model* $\mathcal{M}[G^{\forall}] = \langle W, \{R_i[G^{\forall}] \mid i \in A\}, V \rangle$ *(so, domain and valuation remain the same) in which each* $R_i[G^{\forall}]$ *is given by*

$$R_i[G^{\forall}] := \begin{cases} \{(w,v) \in W \times W \mid v \in \bigcup_{\{H \mid i \in H \subseteq_w^{max} G\}} R_H(w)\} & i \in G; \\ R_i & \textit{otherwise}. \end{cases}$$

[4]This 'duality' between the intuitive idea and the semantic counterpart comes from the fact that, in *EL*, the relational model represents not knowledge/beliefs but rather uncertainty. To justify the proposed definition, consider the somehow similar case of the *"everybody in G knows"* modality, which intuitively ascribes a belief φ to a group G iff this belief is held by *every* agent in G. Semantically, this modality is defined as a normal one working on the *union* of the relations of G's members, thus making reachable those worlds that are reachable by *at least one* member of G.

For describing this operation's effects, add an operator $[G!^{\forall}]$ *for each non-empty* $G \subseteq A$, *with its semantic interpretation given by*

$$\mathcal{M}, w \models [G!^{\forall}]\varphi \quad \textit{iff} \quad \mathcal{M}[G^{\forall}], w \models \varphi \qquad \blacktriangleleft$$

Thus, while indistinguishability for agents in $A \setminus G$ is not affected, that for agents in G is such that

$$R_i[G^{\forall}]wv \qquad \text{iff} \qquad R_H wv \text{ for some } H \text{ such that } i \in H \subseteq_w^{max} G.$$

To highlight the relationship between this operation and that for knowledge resolution (Definition 2.6), recall that the latter is given, for agents in G, by

$$R_i[G^{\forall}]wv \qquad \text{iff} \qquad R_G wv.$$

Thus, while knowledge G-resolution makes v an alternative for i from w if and only if it is an alternative for G from w, cautious belief G-resolution makes v an alternative for i from w if and only if it is an alternative from w for some maximally consistent subset of G to which i belongs. Note that, in knowledge models, the two operations coincide. This is because, when the relations are reflexive, every G has exactly one maximal consistent subgroup at any w: the group G itself.

Preserving properties? The cautious belief resolution does not create inconsistent agents, as it preserves seriality.

Proposition 3.1. *For any group* $\varnothing \subset G \subseteq A$*: if* $\mathcal{M}$ *is a serial model, then so is* $\mathcal{M}[G^{\forall}]$.

Proof. Take any serial model $\mathcal{M} = \langle W, \{R_i \mid i \in A\}, V\rangle$ and any $\varnothing \subset G \subseteq A$. Take any $i \in A$. On the one hand, if $i \notin G$, then $R_i[G^{\forall}]$ is R_i and hence serial. On the other hand, if $i \in G$, then $R_i[G^{\forall}] = \{(w, v) \in W \times W \mid v \in \bigcup_{\{H \mid i \in H \subseteq_w^{max} G\}} R_H(w)\}$. Since R_i is serial, for every $w \in W$ there is at least one consistent subgroup of G that i belongs to (namely, $\{i\}$), which can then be extended into a maximally consistent $H \subseteq_w^{max} G$ to which i belongs. Since such an H is consistent at w, we have $R_H(w) \neq \varnothing$. Thus, for every $w \in W$ we have that $\bigcup_{\{H \mid i \in H \subseteq_w^{max} G\}} R_H(w) \neq \varnothing$; then, $R_i[G^{\forall}]$ is serial. Hence, every $R_i[G^{\forall}]$ is serial, and thus so is $\mathcal{M}[G^{\forall}]$. $\square$

However, consistency is not the only property beliefs require, and thus not the only property a belief communication operation should preserve.

Example 3.1. *First, consider the belief model* $\mathcal{M}_2$ *(below left), representing consistent and fully introspective agents. Its cautious* $\{a, b\}$*-resolved update* $\mathcal{M}_2[\{a, b\}^{\forall}]$ *(below right) is not Euclidean: e.g., both* $R_a wu_2$ *and* $R_a wu_1$ *hold, and yet* $R_a u_2 u_1$ *fails.*

$\mathcal{M}_2$

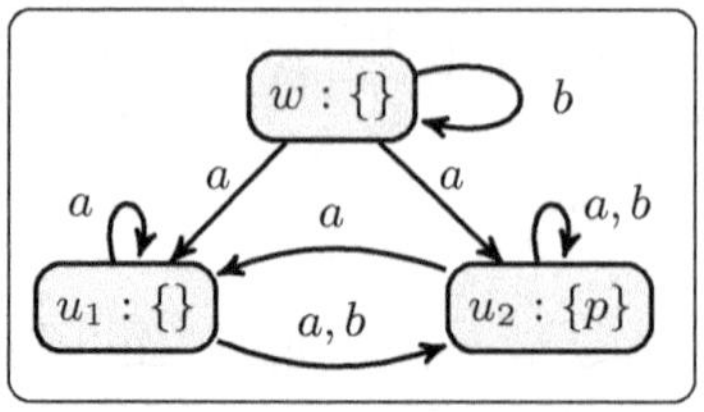

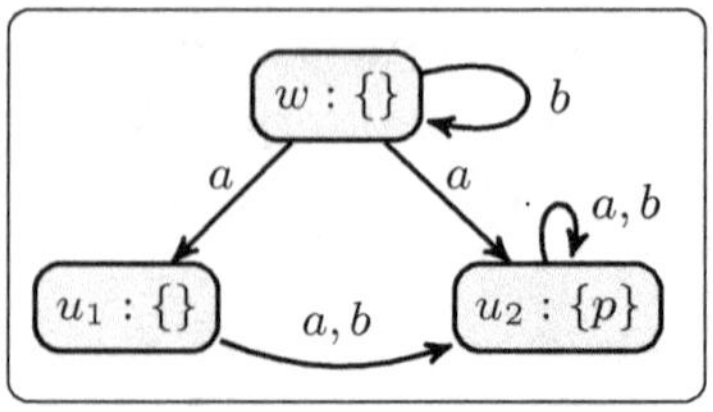

$\mathcal{M}_2[\{a,b\}^{\forall}]$

Thus, through the action, a loses negative introspection. Indeed, after cautious belief resolution, a does not believe p; nevertheless, not only she does not believe this: she actually believes that she believes in p. In symbols,

$$\mathcal{M}_2[\{a,b\}^{\forall}], w \models \neg \mathrm{B}_a p \wedge \neg \mathrm{B}_a \neg \mathrm{B}_a p \wedge \mathrm{B}_a \mathrm{B}_a p.$$

Then, consider the belief model $\mathcal{M}_3$ (below left). Its cautious $\{a,b,c\}$-resolved update $\mathcal{M}_3[\{a,b,c\}^{\forall}]$ (below right) is not transitive (neither Euclidean): e.g., both $R_a w u_1$ and $R_a u_1 u_2$ hold, and yet $R_a w u_2$ fails (e.g., $R_a u_1 u_2$ and $R_a u_1 u_1$ hold, yet $R_a u_2 u_1$ fails).

$\mathcal{M}_3$

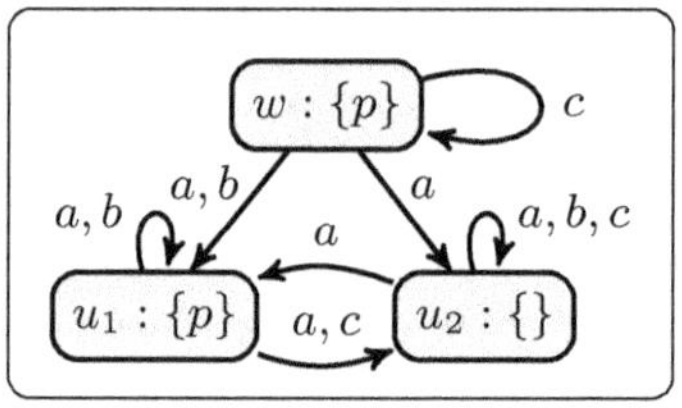

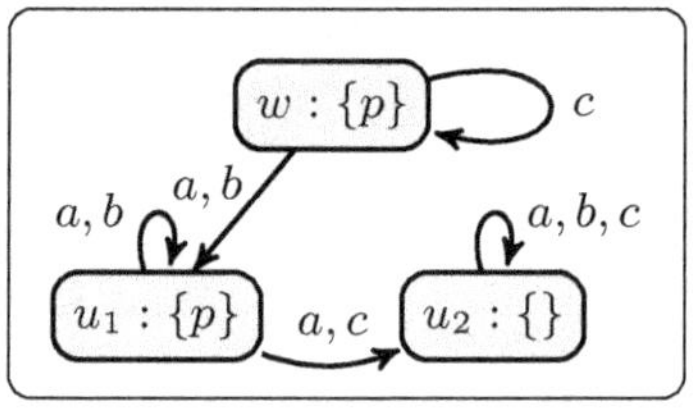

$\mathcal{M}_3[\{a,b,c\}^{\forall}]$

Thus, through the action, a loses positive (and negative) introspection. Indeed, after cautious communication, a believes p; nevertheless, she does not believe that she believes it. In symbols,

$$\mathcal{M}_3[\{a,b,c\}^{\forall}], w \models \mathrm{B}_a p \wedge \neg \mathrm{B}_a \mathrm{B}_a p. \qquad \blacktriangleleft$$

This example shows that the cautious belief G-resolution, on its own, is not adequate for the specific purposes of this paper. Still, it might be useful for data operations that only require to preserve seriality. More importantly, and as it will be discussed, it can be used in tandem with the *sharing* operation that will be introduced in Section 4. For these reasons, it makes sense to look at its profile (expressivity and axiomatisation, discussed below—for technical reasons—in the opposite order).

Axiomatisation For axiomatising a modality whose semantic interpretation relies on a model operation, a useful strategy is the use of *reduction/recursion axioms*: valid formulas and validity-preserving rules defining a recursive translation that takes a formula with model-changing modalities (a formula in the 'dynamic' language) and returns one without them (a formula in the initial 'static' language). To prove soundness when using this strategy, it is enough to show that these axioms and rules are sound (which, in particular, shows also that a formula and its translation are semantically equivalent). Completeness then follows automatically: the recursion axioms make a formula and its translation provably equivalent, and thus one can rely on the completeness of the axiom system for the 'static' language. The reader is referred to van Ditmarsch et al. [45, Chapter 7] and Wang and Cao [48] for an extensive explanation of this technique.

[1] use this strategy for axiomatising $\mathcal{L}_{\mathrm{D},[G!]}$, the language that extends $\mathcal{L}_{\mathrm{D}}$ with the knowledge resolution modality $[G!]$ of Definition 2.6.[5] Indeed, over models with equivalence relations, a sound and complete derivation system consists of the axioms and rules in Table 1, axioms characterising the fact that the relations R_G are of equivalence (i.e., the well-known axioms T, $\mathsf{4}$ and $\mathsf{5}$ for D_G) and the following reduction axioms:

$$\vdash [G!]p \leftrightarrow p \qquad \vdash [G!]\,\mathrm{D}_H\,\varphi \leftrightarrow \mathrm{D}_{G\cup H}[G!]\varphi \quad \text{for } G \cap H \neq \varnothing$$

$$\vdash [G!]\neg\varphi \leftrightarrow \neg[G!]\varphi \qquad \vdash [G!]\,\mathrm{D}_H\,\varphi \leftrightarrow \mathrm{D}_H[G!]\varphi \quad \text{for } G \cap H = \varnothing$$

$$\vdash [G!](\varphi \wedge \psi) \leftrightarrow ([G!]\varphi \wedge [G!]\psi) \qquad \text{If } \vdash \varphi \text{ then } \vdash [G!]\varphi$$

Below it is shown that the same strategy can be used for axiomatising $\mathcal{L}_{\mathrm{D},[G!^{\forall}]}$, the language extending $\mathcal{L}_{\mathrm{D}}$ with the cautious resolution operator. Here are the crucial translation-defining validities (recall: $max_G(H)$ expresses that group H is maximally consistent relative to a group G).

Proposition 3.2 (Reduction principles for $[G!^{\forall}]$). *For any $\varnothing \subset G, H \subseteq A$ and any $\varphi, \psi \in \mathcal{L}_{\mathrm{D},[G!^{\forall}]}$,*

(1) $\models [G!^{\forall}]p \leftrightarrow p$

(2) $\models [G!^{\forall}]\neg\varphi \leftrightarrow \neg[G!^{\forall}]\varphi$

(3) $\models [G!^{\forall}](\varphi \wedge \psi) \leftrightarrow ([G!^{\forall}]\varphi \wedge [G!^{\forall}]\psi)$

(4) $\models [G!^{\forall}]\,\mathrm{D}_H\,\varphi \leftrightarrow \Big(\bigwedge_{F_1 \subseteq G \cap H} \bigwedge_{F_2 \subseteq G} (max_G(F_1 \cup F_2) \rightarrow \mathrm{D}_{F_2 \cup H}[G!^{\forall}]\varphi)\Big)$ *when $G \cap H \neq \varnothing$.*

(5) $\models [G!^{\forall}]\,\mathrm{D}_H\,\varphi \leftrightarrow \mathrm{D}_H[G!^{\forall}]\varphi$ *when $G \cap H = \varnothing$*

[5] In fact, they axiomatise the language $\mathcal{L}_{\Box,\mathrm{D},[G!]}$ (Theorem 34), called $\mathcal{RD}$ in the original paper. Still, as discussed, $\mathcal{L}_{\Box,\mathrm{D}}$ and $\mathcal{L}_{\mathrm{D}}$ are equally expressive, and thus so are $\mathcal{L}_{\Box,\mathrm{D},[G!]}$ and $\mathcal{L}_{\mathrm{D},[G!]}$.

Proof. See the appendix. □

A reduction axiom for the single-agent 'belief' modality $\Box_i$ and $[G!^{\forall}]$ (in the style of those on Equation 1 for knowledge resolution) can be obtained as particular cases of the validities in Item *(4)* and Item *(5)* above.[6] For this, simply rely on the validity of $\Box_i\varphi \leftrightarrow \mathrm{D}_{\{i\}}\,\varphi$. Then,

$$\models [G!^{\forall}]\Box_i\varphi \leftrightarrow \textstyle\bigwedge_{i\in F\subseteq G}\left(max_G(F) \to \mathrm{D}_F[G!^{\forall}]\varphi\right) \quad \text{for } i \in G$$
$$\models [G!^{\forall}]\Box_i\varphi \leftrightarrow \Box_i[G!^{\forall}]\varphi \quad \text{for } i \notin G$$

Thus, an agent in G gets to 'believe' φ after the action iff, before the action, every maximally consistent subset of G she was part of 'believed', distributively, that the action would make φ true.[7] Agents not in G need a stronger requirement: they need to 'believe' before, on their own, that the action would make φ true.

Theorem 3. *The axiom system consisting of the axioms and rules on Table 1, together with axioms for the reduction principles in Proposition 3.2 and the generalisation rule for* $[G!^{\forall}]$*,*

$$\text{if } \vdash \varphi, \text{ then } \vdash [G!^{\forall}]\varphi, \qquad (\mathsf{G}_{[G!^{\forall}]})$$

is sound and strongly complete for $\mathcal{L}_{\mathrm{D},[G!^{\forall}]}$ *w.r.t. the class of all pointed models.*

Proof. The soundness of most axioms and rules has been already discussed (, for those in Table 1; Proposition 3.2 for the reduction principles). For $(\mathsf{G}_{[G!^{\forall}]})$, its soundness in arbitrary models is straightforward. This makes any formula in $\mathcal{L}_{\mathrm{D},[G!^{\forall}]}$ and its translation *semantically* equivalent.

For completeness, note how the rule of preservation of equivalents under the scope of $[G!^{\forall}]$,

$$\text{if } \vdash \varphi \leftrightarrow \psi \text{ then } \vdash [G!^{\forall}]\varphi \leftrightarrow [G!^{\forall}]\psi, \qquad (\mathsf{RE}_{[G!^{\forall}]})$$

is derivable (using PR, $\mathsf{G}_{[G!^{\forall}]}$ and axioms for Items *(2)* and *(3)* of Proposition 3.2).[8] This and the reduction principles make any formula in $\mathcal{L}_{\mathrm{D},[G!^{\forall}]}$ and its translation

[6]We use "$\Box_i$" instead of "B_i" (plus quotation marks on "believe") because the models under discussion are not belief models.

[7]The right-hand side of this equivalence and the way cautious distributed belief is defined (Page 12) are quite similar, but there is an important difference: while the latter 'quantifies universally' over all maximally consistent subgroups of G, the former 'quantifies universally' over all maximally consistent subgroups of G *to which i belongs*.

[8]In [ ] it is shown that for Public Announcement Logic (PAL), the generalisation rule along with the reduction axioms are *not* enough to derive the rule of preservation of equivalents. That is however related to the precondition of the announcement formulas in PAL. For the resolution operators discussed here, the derivation is straightforward.

provably equivalent. In this way, any valid formula in $\mathcal{L}_{\mathrm{D},[G!^{\forall}]}$ can be translated, using the axiom system, into a semantically equivalent (and hence valid) one in $\mathcal{L}_{\mathrm{D}}$, whose derivability in the system is guaranteed by the completeness of the axioms and rules for $\mathcal{L}_{\mathrm{D}}$ (Table 1). Again, an extensive explanation of this technique can be found in van Ditmarsch et al. [45, Chapter 7] and Wang and Cao [48]. □

Expressivity Note how the existence of reduction axioms for $[G!^{\forall}]$ on $\mathcal{L}_{\mathrm{D}}$ implies that $\mathcal{L}_{\mathrm{D},[G!^{\forall}]}$ is invariant under collective bisimilarity.

Theorem 4 ($\rightleftarrows_{\mathrm{D}}$ implies $\leftrightsquigarrow_{\mathcal{L}_{\mathrm{D},[G!^{\forall}]}}$). *Let* $(\mathcal{M}, w)$ *and* $(\mathcal{M}', w')$ *be two pointed models. If* $(\mathcal{M}, w) \rightleftarrows_{\mathrm{D}} (\mathcal{M}', w')$, *then* $(\mathcal{M}, w) \leftrightsquigarrow_{\mathcal{L}_{\mathrm{D},[G!^{\forall}]}} (\mathcal{M}', w')$.

Proof. The crucial point is that "$\leftrightsquigarrow_{\mathcal{L}_{\mathrm{D}}}$" is equivalent to "$\leftrightsquigarrow_{\mathcal{L}_{\mathrm{D},[G!^{\forall}]}}$". From right to left this is immediate: two models agreeing in all formulas in $\mathcal{L}_{\mathrm{D},[G!^{\forall}]}$ also agree in all formulas in its subset $\mathcal{L}_{\mathrm{D}}$. From left to right, agreeing in all formulas in $\mathcal{L}_{\mathrm{D}}$ implies agreement in all formulas in $\mathcal{L}_{\mathrm{D},[G!^{\forall}]}$, as any formula of the latter is semantically equivalent to some formula in the former, witness the translation induced by the reduction axioms. From this and Theorem 1, the result follows. □

3.2 Patching cautious belief resolution

While cautious resolution preserves seriality (consistency), it fails to preserve the other properties for a relation representing beliefs: transitivity (positive introspection) and Euclidicity (negative introspection). To deal with this, one can look at the natural ways for a relation R to (re)gain these properties: the transitive and Euclidean closures (i.e., the smallest transitive/Euclidean relation containing the original), respectively denoted by R^+ and R^E.[9] Still, note: in general, Euclidean closure does not preserve transitivity. Thus, the order of application of these operations matters. Since transitive closure preserves Euclidicity, our approach starts with the cautious update, applying then the Euclidean closure and finally the transitive closure. (Formal definitions of these closure operations as well as their technical details can be found in the appendix.)

Definition 3.4 (Patched cautious belief G-resolution). *Let* $\mathcal{M} = \langle W, \{R_i \mid i \in A\}, V\rangle$ *be a model; take* $\varnothing \subset G \subseteq A$. *The* patched cautious G-resolved update *of*

[9]This idea does not work for regaining seriality: there is not such a thing as 'the serial closure' because, in general, there is no single smallest serial relation containing the original. One could still work by quantifying over the different *minimal* ways a relation can be made serial (add only one outgoing i-edge to each world lacking it). Nevertheless, this would still involve a good amount of arbitrariness: *any* world in the model is a candidate target for the to-be-added edge.

$\mathcal{M}$ is the model $\mathcal{M}[G^{\odot}] = \langle W, \{R_i[G^{\odot}] \mid i \in A\}, V\rangle$ in which each $R_i[G^{\odot}]$ is given by

$$R_i[G^{\odot}] := ((R_i[G^{\forall}])^E)^+$$

with $R_i[G^{\forall}]$ as in Definition 3.3. For describing this operation's effects, add an operator $[G!^{\odot}]$ for each $\varnothing \subset G \subseteq A$, with its semantic interpretation given by

$$\mathcal{M}, w \vDash [G!^{\odot}]\varphi \qquad \textit{iff} \qquad \mathcal{M}[G^{\odot}], w \vDash \varphi$$ ◀

Preserving properties This operation preserves belief models.

Proposition 3.3. *For any group $\varnothing \subset G \subseteq A$: if $\mathcal{M}$ is a belief model, then so is $\mathcal{M}[G^{\odot}]$.*

Proof. Each relation in $\mathcal{M}[G^{\odot}]$ is given by $R_i[G^{\odot}] := ((R_i[G^{\forall}])^E)^+$. For seriality recall that, if $\mathcal{M}$ is serial, so is $R_i[G^{\forall}]$ (Proposition 3.1). Hence, so are $(R_i[G^{\forall}])^E$ and then $((R_i[G^{\forall}])^E)^+$, as the closure operations do not erase edges and the domain remains the same. Transitivity follows from the construction's last step, transitive closure. For Euclidicity, $(R_i[G^{\forall}])^E$ is Euclidean; thus (by Proposition A.1 in the appendix), so is $((R_i[G^{\forall}])^E)^+$. □

Effects of the operation The just defined operation preserves the relational properties of belief (and thus preserves consistent as well as positively and negatively introspective agents). Now let's look at its effects. This is useful/important as, while the first step in the model construction process can only remove edges (so the agent can lose uncertainty), both Euclidean and transitive closure might add them (so the agent can gain uncertainty).

First, there are belief cases in which the edges added by the closure operations still leave the agent with new information.

Example 3.2. *Consider the belief model $\mathcal{M}_4$ below on the left. In w, agent a considers u_3 (where p and q are both true), u_1 (p is true and q false) and u_2 (both p and q are false). Note how a is uncertain about both atoms: $\mathcal{M}_4, w \vDash (\neg \mathrm{B}_a\, p \wedge \neg \mathrm{B}_a \neg p) \wedge (\neg \mathrm{B}_a\, q \wedge \neg \mathrm{B}_a \neg q)$. Yet, b considers only u_1 (so $\mathcal{M}_4, w \vDash \mathrm{B}_b\, p \wedge \mathrm{B}_b \neg q)$) and c considers only u_3 (so $\mathcal{M}_4, w \vDash \mathrm{B}_c\, p \wedge \mathrm{B}_c\, q)$). Thus, a is consistent with both b and c, but b and c are mutually inconsistent.*

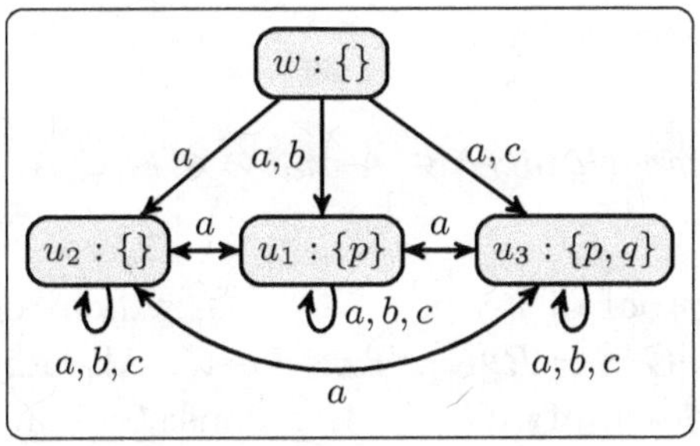

$\mathcal{M}_4$

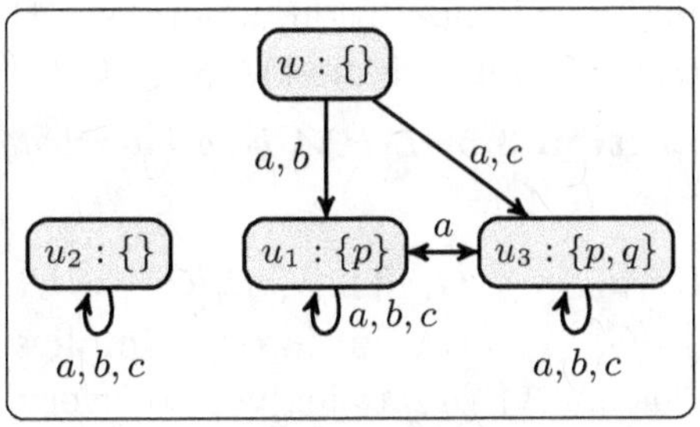

$\mathcal{M}_4[\{a,b,c\}^{\odot}]$

Now, observe the patched cautious update $\mathcal{M}_4[\{a,b,c\}^{\odot}]$ (above right). At w, agent a is consistent with both b and c, but b and c are mutually inconsistent; thus, the worlds she considers from w are now $\{u_1, u_3\}$. Hence, while a maintains uncertainty about q, she now believes p. In symbols, $\mathcal{M}_4[\{a,b,c\}^{\odot}], w \models (\neg \mathrm{B}_a\, q \wedge \neg \mathrm{B}_a \neg q) \wedge \mathrm{B}_a\, p$. ◀

Then, there are cases in which the additional work the closure operations makes an agent 'forget'/'ignore' the information she received.

Example 3.3. *Consider the non-Euclidean model $\mathcal{M}_2[\{a,b\}^{\forall}]$ below on the left, which resulted from applying cautious belief $\{a,b\}$-resolution to $\mathcal{M}_2$ (both from Example 3.1). Completing the patched cautious $\{a,b\}$-resolved update produces the model $\mathcal{M}_2[\{a,b\}^{\odot}]$ below on the right, which is identical to the original $\mathcal{M}_2$.*

$\mathcal{M}_2[\{a,b\}^{\forall}]$

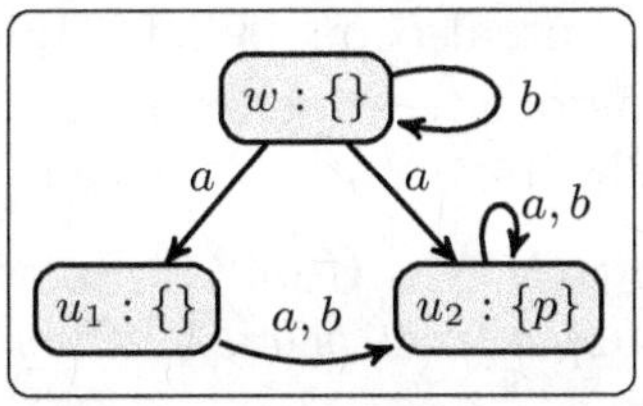

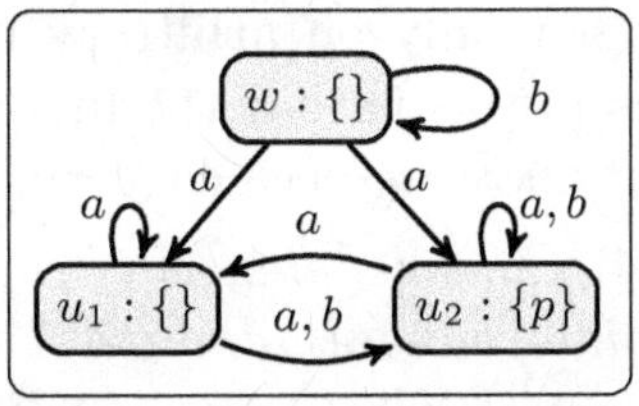

$\mathcal{M}_2[\{a,b\}^{\odot}]$

The same happens when the patched cautious $\{a,b,c\}$-resolved update is applied to $\mathcal{M}_3$ (again, in Example 3.1): $\mathcal{M}_3$ and $\mathcal{M}_3[\{a,b,c\}^{\odot}]$ are the same. ◀

Still, when working with Euclidean and transitive relations, the latter is this operation's 'worst' possible scenario: the agent might not learn anything, but she will not forget either. In relational terms, the agent will not have, after the operation, more uncertainty than what she initially had.[10]

Proposition 3.4. *Let $\mathcal{M}$ be a transitive and Euclidean model; take $i \in A$ and $\varnothing \subset G \subseteq A$. Then, $R_i[G^{\odot}] \subseteq R_i$.*

Proof. See the appendix. □

[10]In non-Euclidean/non-transitive models, the operations will clearly add brand new edges.

Finally, this operation generalises the knowledge case of Ågotnes and Wáng [ ] (Definition 2.6 in this text).

Proposition 3.5. *Let $\mathcal{M}$ be a knowledge model; take $i \in A$ and $\varnothing \subset G \subseteq A$. Then, $R_i[G^{\odot}] = R_i[G]$.*

Proof. Take $i \notin G$. Then, $R_i[G^{\odot}] = R_i$ (proof of Proposition 3.4), which together with $R_i[G] = R_i$ (Definition 2.6) implies $R_i[G^{\odot}] = R_i[G]$. Take $i \in G$. All individual relations in $\mathcal{M}$ are reflexive, so every non-empty $G \subseteq A$ is consistent at every world, and thus it has, at every world, only one maximally consistent subset: G itself. Hence, the first step in the patched cautious update, $R_i[G^{\forall}] = \{(w,v) \in W \times W \mid v \in \bigcup_{\{H \mid i \in H \subseteq^{max}_{w} G\}} R_H(w)\}$, becomes $R_i[G^{\forall}] = \{(w,v) \in W \times W \mid v \in R_G(w)\} = R_G$. Since all individual relations in $\mathcal{M}$ are Euclidean and transitive, so is the intersection R_G, and thus so is $R_i[G^{\forall}]$. Hence, the second and third steps in the patched cautious update become, respectively, $(R_i[G^{\forall}])^E = R_i[G^{\forall}]$ and thus $((R_i[G^{\forall}])^E)^+ = (R_i[G^{\forall}])^E$. Hence, $R_i[G^{\odot}] = R_G$, which together with $R_i[G] = R_G$ (Definition 2.6) implies $R_i[G^{\odot}] = R_i[G]$. □

Having established the effects of a patched cautious belief resolution, it is time to look at its modality's profile.

Expressivity Unlike the $[G!^{\forall}]$ case, adding $[G!^{\odot}]$ to $\mathcal{L}_{\mathrm{D}}$ increases the expressivity of the latter, not only on arbitrary models, but also on belief models. The crucial point behind this is that the modality's associated operation, patched cautious resolution, does not preserve collective bisimilarity.

Example 3.4. *Consider the belief pointed models $\mathcal{M}_2$ (from Example 3.1, with the evaluation point changed) and $\mathcal{M}_2'$ in the upper row below. They are collective bisimilar (witness $\{(u_2, u_2'), (u_1, u_1')\}$).*

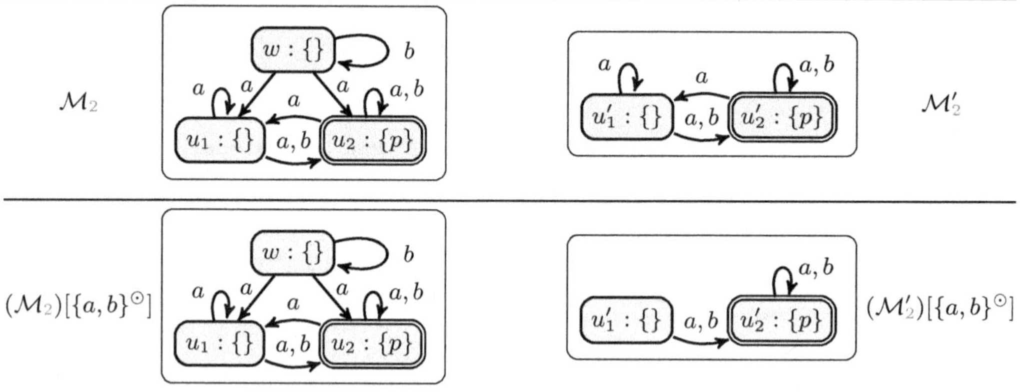

The patched cautious belief $\{a, b\}$*-resolution produces, respectively, the pointed models in the lower row. Note how they are not collectively bisimilar: the* $\{a\}$*-edge from* u_2 *to a* $\neg p$*-world in* $\mathcal{M}_2$ *does not have a matching* $\{a\}$*-transition from* u_2' *in* $\mathcal{M}_2'$. ◀

These examples serve as witnesses for the following proposition.

Proposition 3.6. *On arbitrary models as well as on belief models, "*$\leftrightarrows_{\mathrm{D}}$*" does not imply "*$\leftrightsquigarrow_{\mathcal{L}_{\mathrm{D},[G!\odot]}}$*".*

Proof. The initial pointed models in Example 3.4, $(\mathcal{M}_2, u_2)$ and $(\mathcal{M}_2', u_2')$, are collectively bisimilar, thus $\mathcal{L}_{\mathrm{D}}$-equivalent. Yet, after the update, the models are not $\mathcal{L}_{\mathrm{D}}$-equivalent: they can be distinguished by the $\mathcal{L}_{\mathrm{D}}$-formula $\mathrm{B}_a\, p$, which is false in the first but true in the second. Thus, the initial pointed models can be distinguished by the $\mathcal{L}_{\mathrm{D},[G!\odot]}$-formula $[\{a, b\}!^{\odot}]\,\mathrm{B}_a\, p$. □

Corollary 3.1. *On arbitrary models as well as on belief ones,* $\mathcal{L}_{\mathrm{D},[G!\odot]}$ *is strictly more expressive than* $\mathcal{L}_{\mathrm{D}}$. ◀

It is worthwhile to look closer at the structures in Example 3.4, to better understand the reason why collective bisimilarity is not preserved. First, models $\mathcal{M}_2$ and $\mathcal{M}_2'$ are different: while the submodels given by $\{u_1, u_2\}$ and $\{u_1', u_2'\}$ are identical,[11] the former contains an additional world, w. Still, w is not reachable from the model's *evaluation point* by any finite sequence of group-transitions. Thus, both *pointed models* are, for collective bisimilarity (and thus, for $\mathcal{L}_{\mathrm{D}}$ [Theorem 1]), the same. However, the presence of w influences the behaviour of the patched cautious update: when it is present (model $\mathcal{M}_2$), the a-arrow from u_2 to u_1 is preserved (model $(\mathcal{M}_2)[\{a, b\}^{\odot}]$); when it is *not* (model $\mathcal{M}_2'$), the mentioned arrow is removed (model $(\mathcal{M}_2')[\{a, b\}^{\odot}]$). This is important because this difference concerns whether a world is reachable from the evaluation point by a finite sequence of group-steps (namely, a single $\{a\}$-step); thus, the difference concerns collective bisimilarity. The 'culprit' for this behaviour is the Euclidean closure operation and, in particular, its use of 'backward' transitions (Definition A.1): the worlds a given v can reach after the operation depends not only on the worlds it could reach before but also on the worlds that could reach it. This is what breaks collective bisimilarity.

There is also a more conceptual side to this. Individual and distributed beliefs are defined by what can be reached. Thus, it is odd that *the way they change* is affected by worlds that are unreachable. This can be made precise by focussing on $\mathcal{M}_2$ and taking w as the evaluation point. Here, a is uncertain about p and, moreover, mistakenly believes b believes that p is true (i.e., $(\mathcal{M}_2, w) \vDash (\neg\,\mathrm{B}_a\, p \wedge \neg\,\mathrm{B}_a \neg p) \wedge (\mathrm{B}_a\,\mathrm{B}_b\, p \wedge \mathrm{B}_b \neg p)$), and also mistakenly believes she is consistent with

[11] More technically, the submodels generated respectively by $\{u_2\}$ and $\{u_2'\}$ are exactly the same.

b (i.e., $(\mathcal{M}_2, w) \models \mathrm{B}_a \neg \mathrm{D}_{\{a,b\}} \bot \wedge \mathrm{D}_{\{a,b\}} \bot$). Thus, intuitively, she should believe that, after the action, she will believe p. However, this is not the case: $(\mathcal{M}_2, w) \not\models \mathrm{B}_a[\{a, b\}!^{\odot}]\, \mathrm{B}_a\, p$. After noticing that the intuition actually holds when the action is only the cautious belief resolution (i.e., $(\mathcal{M}_2, w) \models \mathrm{B}_a[\{a, b\}!^{\forall}]\, \mathrm{B}_a\, p$), one is inclined to think that the closure operations, while required for technical reasons, are not conceptually appropriate.

Axiomatisation Above, it has been argued that the patched cautious resolution might not be conceptually appropriate. Still, for 'technical completeness', it is worthwhile to discuss, briefly, the axiomatisation of its associated modality.

Since $\mathcal{L}_{\mathrm{D},[G!^{\odot}]}$ is more expressive than $\mathcal{L}_{\mathrm{D}}$, there are no reduction axioms from the former to the latter. In such cases, a typical axiomatisation strategy consists on looking for additional modal operators that, when added to the basic language, make it expressive enough to allow the existence of reduction axioms. To complete the task, one should also axiomatise the newly added modalities. An example of this strategy is the case of the language $\mathcal{L}_{\mathrm{K},C}$ (including knowledge operators K_i for individual agents $i \in A$ and common knowledge operators C_G for groups $G \subseteq A$) and its extension with the public announcement operator $[\varphi!]$. It has been shown [ ] that $\mathcal{L}_{\mathrm{K},C,[\varphi!]}$ is strictly more expressive than $\mathcal{L}_{\mathrm{K},C}$, so no reduction axioms from the former to the latter exists. However, adding the operator for *relativized common knowledge* (C_G^{φ}) helps: there are reduction axioms from $\mathcal{L}_{\mathrm{K},C,C^{\varphi},[\varphi!]}$ to $\mathcal{L}_{\mathrm{K},C,C^{\varphi}}$ [ ].

Such an approach is possible here, though not in a straightforward way. To see this, observe that the crucial issue lies in finding a reduction for $[G!^{\odot}]\, \mathrm{D}_H\, \varphi$—a formula whose behaviour is essentially that of a normal universal modality semantically interpreted over the relation $\bigcap_{i \in H}((R_i[G^{\forall}])^E)^+$. What is needed is, then, a way of expressing that, *prior to the operation*, the formula φ holds at every world reachable via this relation. While the relation involves operations definable within $\mathcal{L}_{\mathrm{D}}$ (e.g., union and sequential composition), it also includes non-definable ones—such as converse, transitive closure and reflexive & transitive closure—hence the increase in expressivity. Furthermore, it involves intersections *after* the these operations have been applied. Therefore, simply adding modalities for the Euclidean and transitive closure of the individual relations R_i (the strategy used in to axiomatise, *separately*, model operations for Euclidean and transitive closures) is insufficient. Identifying a extension suitable for axiomatising $\mathcal{L}_{\mathrm{D},[G!^{\odot}]}$ is an interesting technical task left for future work.[12]

[12]An axiomatization without reduction axioms, in the style of Wang and Cao [, Section 4] for public announcements—where $[G!^{\odot}]$ is interpreted not via a model operation but rather as a relational transition in a so-called 'supermodel'—would still require reference to converse, transitive closure, and reflexive & transitive closure of individual relations.

4 Belief sharing

4.1 Taking a step back

Recall that, inspired by the knowledge resolution operation [ ], this paper's goal is to provide a model operation representing an action of *belief* resolution, an act through which a group of agents communicate their beliefs to one another. An initial attempt that simply mimics the knowledge resolution strategy turns out to have an important drawback: it does not preserve the consistency of beliefs (technically, it does not preserve seriality; Example 2.1). Thus, Section 3 examined two alternatives. The first, cautious belief resolution (Definition 3.3), preserves consistency (technically, seriality), but it does not preserve the two other properties typically ascribed to beliefs: positive and negative introspection (technically, transitivity and Euclidicity). The second, patched cautious belief resolution (Definition 3.4), preserves the three required properties. Still, as it has been discussed, it is not conceptually adequate.

When looking for a further alternative, it is important to emphasise that, to preserve the discussed properties, the proposed operations 'reject' some of the information the agents receive. This is not needed in the knowledge case: the information each agent has is truthful (technically, the relations are reflexive) and thus no conflict might arise. In other words, it is impossible for two agents to have opposite views on *any* given φ; hence, every agent can 'accept' the incoming information as it is. This is different in the beliefs case: consistent agents might be, as a group, inconsistent. Thus, to preserve consistency, some information must be rejected.

Now, while it is reasonable for an agent to reject *some* of the information she receives, it does not mean she should reject *any* of the provided information. In particular, while she might reject other agents' beliefs about *non-epistemic facts* (i.e. other agents' first-order beliefs), she should not reject other agents' beliefs *about their own beliefs*. Consider agents a and b, with b communicating her beliefs. On the one hand, if a believes p and b communicates $\neg p$, it makes sense for a to *reject* b's information so, afterwards, she continues to believe p. On the other hand, if a believes $\mathrm{B}_b\, p$ and b communicates $\mathrm{B}_b \neg p$, it makes sense for a to *accept* b's information (note: about b's own beliefs) so, afterwards, she now believes $\mathrm{B}_b \neg p$.[13] In other words, while other agents' beliefs about non-epistemic facts might be rejected (this is what the [patched] cautious approach does), other agents' beliefs about their own beliefs should be accepted. As seen in the previous section, however, the patched cautious belief resolution will in some cases completely ignore the incoming

[13] An agent's beliefs might be incorrect but, in the action under discussion, agents are sincere: they communicate what they believe. Moreover: given positive and negative introspection, agents always have a correct perception about their own beliefs.

information about the other agents' beliefs in order to preserve belief properties (Example 3.3).

Thus, different parts of the beliefs an agent communicates might/should be treated differently by the receiving agents. Now, given the nature of the EL semantic structure (in particular, the fact that belief models are transitive), in this setting it is not straightforward to disentangle an agent's first-order and higher-order beliefs. Because of this, the strategy to be followed here will be slightly different. Relying on the earlier observation that the agents' beliefs about their own beliefs should be accepted, the belief resolution action will be split into two parts: first, one through which agents learn what the beliefs of others are (and thus potentially fix their own beliefs on this), and then another through which they resolve their (first-order) beliefs with those of the others. While the first step is not needed in the knowledge case, it is important when beliefs are involved: even if an agent wants to stick to her own first-order beliefs, the resolution action should at least correct her beliefs about *other agents'* beliefs.

4.2 A belief sharing operation

The operation below captures the first part of the discussed process: learning what the sharing agents believe. It requires not just a model, but a *pointed* model, and works by making sure *every* agent in the model correctly identifies the conjecture set held by the sharing agents (those in G) *at the evaluation point.*[14]

Definition 4.1 (Belief G-sharing). *Let $(\mathcal{M}, w)$ be a pointed model with $\mathcal{M} = \langle W, \{R_i \mid i \in A\}, V\rangle$; take $\varnothing \subset G \subseteq A$. The* G-share update *of $\mathcal{M}$ at w is the* model $\mathcal{M}[G^S_w]$ *with $\mathcal{M}[G^S_w] = \langle W, \{R_i[G^S_w] \mid i \in A\}, V\rangle$ (so, domain and valuation remain the same) in which each $R_i[G^S_w]$ is given by*

$$R_i[G^S_w] := \begin{cases} \{(v,u) \in W \times W \mid u \in R_i(w)\} & i \in G; \\ R_i & \textit{otherwise}. \end{cases}$$

For describing the effects of this operation, add to the language an operator $\mathbf{S}_G$ for each non-empty $G \subseteq A$, with its semantic interpretation given by

$$\mathcal{M}, w \vDash \mathbf{S}_G\varphi \qquad \textit{iff} \qquad \mathcal{M}[G^S_w], w \vDash \varphi.$$

[14]Taking the spirit of the previously discussed resolution operations into consideration, one might prefer an action through which *only the sharing agents* get to correctly identify their conjecture sets (defined, e.g., via the action models of). Here, only the 'simpler' operation described above is discussed, to keep the focus on the main idea: an action for learning what others believe.

When G is a singleton $\{a\}$, we use $\mathbf{S}_a$ instead of $\mathbf{S}_{\{a\}}$. The notation $[\![\varphi]\!]_{\mathcal{M}} := \{w \in W \mid \mathcal{M}, w \vDash \varphi\}$ will be useful. ◀

Thus, while the relation for agents not in G remains unaffected, for those in G, it is updated so that, for every world $v \in W$,

$$R_i[G_w^S](v) = R_i(w).$$

In words, in the resulting model, every world will correctly identify the initial conjecture set of each agent in G at the evaluation point w. As a consequence, every agent in the model will correctly identify the beliefs of agents in G.[15]

Here is an example illustrating how this operation works.

Example 4.1. *Consider the pointed belief model $(\mathcal{M}_5, w)$ below on the left. In it, while agent b believes both $\neg p$ and that a believes $\neg p$ (i.e., $\mathcal{M}_5, w \vDash \mathrm{B}_b \neg p \wedge \mathrm{B}_b \mathrm{B}_a \neg p$), in reality p holds and a believes it ($\mathcal{M}_5, w \vDash p \wedge \mathrm{B}_a p$).*

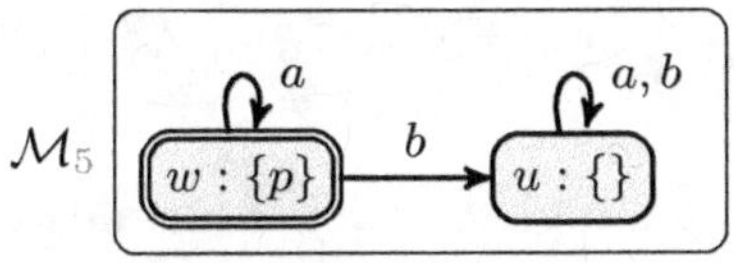

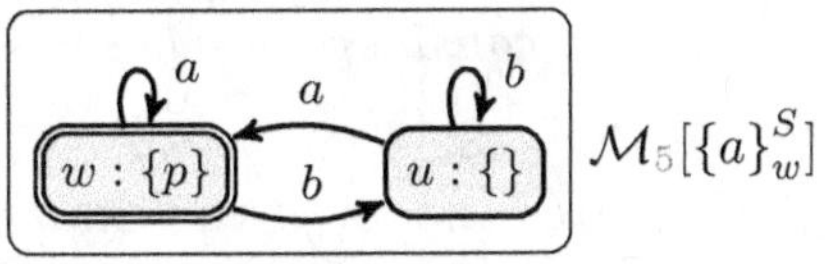

The pointed belief model after an $\{a\}$-shared update (i.e., after all agents learn what a believes) is shown above on the right. In it, b continues to believe $\neg p$, but now correctly believes that a believes p (i.e., $\mathcal{M}_5[\{a\}_w^S], w \vDash \mathrm{B}_b \neg p \wedge \mathrm{B}_b \mathrm{B}_a p$). ◀

Preserving properties Like the patched resolution operation, the sharing operation preserves the relational properties for beliefs.

Proposition 4.1. *For any group $\varnothing \subset G \subseteq A$: if $\mathcal{M}$ is a belief model and w a world in it, then $\mathcal{M}[G_w^S]$ is a belief model.*

Proof. On the one hand, if $i \notin G$, then trivially any property of R_i will hold also of $R_i[G_w^S]$. On the other hand, if $i \in G$, we have the following. First, assume R_i is serial. Then, in particular, there is $u \in W$ such that $R_i wu$, and thus $R_i[G_w^S]vu$ holds for all $v \in W$. Thus, $R_i[G_w^S]$ is serial. Second, $R_i[G_w^S]$ is transitive regardless of the properties of R_i. Indeed, take $x, y, z \in W$ such that $R_i[G_w^S]xy$ and $R_i[G_w^S]yz$. By definition, $R_i[G_w^S]yz$ iff $R_i wz$ iff $R_i[G_w^S]xz$. Third, that $R_i[G_w^S]$ is Euclidean regardless of the properties of R_i can be proved just as in the transitivity case. □

[15] Note that a 'weaker' operation, one correcting the conjecture sets of agents in G *only in worlds that are relevant for some agent* (i.e., only in worlds reachable from w via $(\bigcup_{i \in A} R_i)^+$), might be enough. We have chosen the 'stronger' operation defined above for simplicity.

Still, note: the G-share update does not preserve reflexivity. Thus, this operation is not suitable for knowledge models. The issue is not on first-order information (which, by Proposition 4.2 below, remains unaffected), but rather on higher-order information, as illustrated in the following example.

Example 4.2. *Consider the pointed knowledge model $(\mathcal{M}_6, w)$ (below on the left) and its $\{a\}$-shared update (below on the right).*

Note how $(u, u) \notin R_a[\{a\}^S_w]$, so $\mathcal{M}_6[\{a\}^S_w]$ is not reflexive. This affects the knowledge of the agents: while in the initial pointed model b knows a's knowledge is factive (i.e., $(\mathcal{M}_6, w) \models \mathrm{K}_b(\mathrm{K}_a\, p \to p)$), after the sharing operation this is not the case anymore: now, b considers possible a's knowledge is not factive (i.e., $\mathcal{M}_6[\{a\}^S_w], w \models \widehat{\mathrm{K}}_b\, \neg(\mathrm{K}_a\, p \to p)$). ◀

In the updated model of the example above, b's higher-order information about a changed to correctly identify what a knows (this happens in general; see Proposition 4.5.*(4)* below). However, b does *not* come to know the facts that a knows (this also happens in general; see Proposition 4.2). Thus, the update leads to a situation in which b accepts that a knows p, and yet still considers $\neg p$ possible. From this perspective, the loss of reflexivity is not a 'bug' of the operation, but rather a feature: agents will identify correctly their higher-order beliefs about other agents' beliefs *without thereby accepting the first-order content of those beliefs.* Such a separation is appropriate for belief, where the content of another agent's beliefs may conflict with one's own, but it is not appropriate for knowledge, where accepting that another agent knows φ should also commit one to φ. Thus, the operation should be understood specifically as a belief-learning operation, not as a generalisation of knowledge resolution.

Effects of the operation While our main concern is the way the operation works on belief models, its effects on arbitrary models already gives us a glimpse into its behaviour. Here, two main points are discussed: the operation leaves the agents' first-order beliefs unchanged, and it ensures that all agents correctly identify the beliefs of those in G.

First, *at the point where the update takes place*, all agents' conjecture sets (and thus all combined conjecture sets) remain unchanged.

Proposition 4.2. *Let* $(\mathcal{M}, w)$ *be a pointed model with* $\mathcal{M} = \langle W, \{R_i \mid i \in A\}, V\rangle$*; take* $i \in A$ *and* $\varnothing \subset G \subseteq A$*. Then,*

$$R_i(w) = R_i[G^S_w](w).$$

As a consequence, $R_H(w) = R_H[G^S_w](w)$ *for every* $H \subseteq A$. ◀

Thus, the operation does not affect the agents' first-order beliefs.

Corollary 4.1. *Let* $(\mathcal{M}, w)$ *be a pointed belief model with* $\mathcal{M} = \langle W, \{R_i \mid i \in A\}, V\rangle$*; take* $i \in A$ *and* $G, H \subseteq A$ *as well as any propositional* γ*. Then,*

$$(\mathcal{M}, w) \models \mathrm{B}_i\,\gamma \leftrightarrow \mathbf{S}_G\,\mathrm{B}_i\,\gamma \qquad \textit{and} \qquad (\mathcal{M}, w) \models \mathrm{D}_H\,\gamma \leftrightarrow \mathbf{S}_G\,\mathrm{D}_H\,\gamma.$$

Proof. For the first, $w \in [\![\mathrm{B}_i\,\gamma]\!]_{\mathcal{M}}$ iff $R_i(w) \subseteq [\![\gamma]\!]_{\mathcal{M}}$ iff $R_i[G^S_w](w) \subseteq [\![\gamma]\!]_{\mathcal{M}}$ (Proposition 4.2) iff $R_i[G^S_w](w) \subseteq [\![\gamma]\!]_{\mathcal{M}[G^S_w]}$ (as γ is propositional) iff $w \in [\![\mathrm{B}_i\,\gamma]\!]_{\mathcal{M}[G^S_w]}$ iff $w \in [\![\mathbf{S}_G\,\mathrm{B}_i\,\gamma]\!]_{\mathcal{M}}$. The argument for the second is analogous. □

Second, after a G-sharing update, the conjecture sets of all agents in G are the same in every point in the model, and the same as in the initial pointed model.

Proposition 4.3. *Let* $(\mathcal{M}, w)$ *be a pointed model with* $\mathcal{M} = \langle W, \{R_i \mid i \in A\}, V\rangle$*; take* $i \in G \subseteq A$*. Then, for any* $v_1, v_2 \in W$*:*

$$R_i[G^S_w](v_1) = R_i[G^S_w](v_2) = R_i(w).$$

As a consequence, $R_H[G^S_w](v_1) = R_H[G^S_w](v_2)$ *for every* $H \subseteq G$. ◀

Since the conjecture sets of agents in the sharing group are global in the updated model, so are their (distributed) beliefs.

Corollary 4.2. *Let* $(\mathcal{M}, w)$ *be a pointed model with* $\mathcal{M} = \langle W, \{R_i \mid i \in A\}, V\rangle$*; take* $H \subseteq G \subseteq A$ *and* $\varphi \in \mathcal{L}_{\mathrm{D},\mathbf{S}_G}$*. Then,*

$$(\mathcal{M}[G^S_w], v) \models \mathrm{D}_H\,\varphi \textit{ for all } v \in W \qquad \textit{or} \qquad (\mathcal{M}[G^S_w], v) \not\models \mathrm{D}_H\,\varphi \textit{ for all } v \in W.$$

Proof. Straightforward from Proposition 4.3 as, from any $v \in W$, the initial D_H-step takes us to the same set. □

4.3 Sharing in belief models

Belief models have a particular feature that makes the G-sharing operation useful for our purposes: at any given evaluation point, the relation for any agent (and thus for any group) *preserves* the conjecture set.

Proposition 4.4. *Let* $(\mathcal{M}, w)$ *be a pointed transitive and Euclidean model with* $\mathcal{M} = \langle W, \{R_i \mid i \in A\}, V\rangle$*; take* $i \in A$ *and* $\varnothing \subset G \subseteq A$*. Then, for any* $u \in W$*,*

(1) *$R_i wu$ implies $R_i(w) = R_i(u)$;*

(2) *$R_G wu$ implies $R_G(w) = R_G(u)$.*

Proof. For Item *(1)*, while $R_i(w) \subseteq R_i(u)$ is guaranteed by Euclidicity, $R_i(u) \subseteq R_i(w)$ is guaranteed by transitivity. Item *(2)* follows from the previous, as intersection preserves transitivity and Euclidicity (i.e., R_G is transitive and Euclidean in transitive and Euclidean models). □

With this, here are the main effects of the G-sharing operation in belief models.

Proposition 4.5. *Take any $\varnothing \subset G \subseteq A$.*

(1) *The G-sharing operation does not affect the agents' (or groups of agents') first-order beliefs. In symbols, $\models \mathrm{D}_H\,\gamma \leftrightarrow \mathbf{S}_G\,\mathrm{D}_H\,\gamma$ for any $H \subseteq A$ and any propositional formula γ,*

(2) *The G-sharing operation does not affect the* self *high-order beliefs of any agents. In symbols, $(\mathcal{M}, w) \models \mathrm{B}_i\,\chi \leftrightarrow \mathbf{S}_G\,\mathrm{B}_i\,\chi$ for any $i \in A$ and any $\chi \in \mathcal{L}_{\mathrm{D}}$ whose modalities refer only to agent i.*

(3) *After a G-sharing operation,* every agent in the model *will correctly identify the* self *beliefs of each agent $i \in G$. In symbols, $(\mathcal{M}, w) \models \mathrm{B}_i\,\chi \rightarrow \mathbf{S}_G\,\mathrm{B}_j\,\mathrm{B}_i\,\chi$ for any $j \in A$, any $i \in G$ and any $\chi \in \mathcal{L}_{\mathrm{D}}$ whose modalities refer only to agent i.*

(4) *After a G-sharing operation,* every agent in the model *will correctly identify the updated beliefs of agents in G (and of subgroups of G). In symbols, $\models \mathbf{S}_G(\mathrm{D}_H\,\varphi \leftrightarrow \mathrm{B}_i\,\mathrm{D}_H\,\varphi)$ for any $i \in A$, any $H \subseteq G$ and any $\varphi \in \mathcal{L}_{\mathrm{D},\mathbf{S}_G}$.*

Proof.

(1) This is the second part of Corollary 4.1.

(2) Take any pointed model $(\mathcal{M}, w)$ and agent $i \in A$. The i-paths starting at the point of the update operation are the same before and after the operation. To see this, simply note that for $i \in G$ and $u \in R_i(w)$ and $v \in W$, by Proposition 4.4, Proposition 4.2, and Proposition 4.3, $R_i(w) = R_i(u) = R_i[G^S_w](w) = R_i[G^S_w](v)$. For $i \in A \setminus G$, the i-paths are unaffected by the operation. Since the i-paths are the same before and after the operation, any $\mathcal{L}_{\mathrm{D}}$-formulas contain only the B_i-modality will be satisfied after the operation if and only if they are satisfied before the operation.

(3) Take $j \in A$ and $i \in G$. By the previous point, $(\mathcal{M}, w) \models \mathrm{B}_i\,\varphi \leftrightarrow \mathbf{S}_G\,\mathrm{B}_i\,\varphi$; by Corollary 4.2, $(\mathcal{M}, w) \models \mathbf{S}_G\,\mathrm{B}_i\,\varphi \rightarrow \mathbf{S}_G\,\mathrm{B}_j\,\mathrm{B}_i\,\varphi$. Hence the required $(\mathcal{M}, w) \models \mathrm{B}_i\,\varphi \rightarrow \mathbf{S}_G\,\mathrm{B}_j\,\mathrm{B}_i\,\varphi$.

(4) By Corollary 4.2, if $H \subseteq G$, then for any pointed model $(\mathcal{M}, w)$ and any $i \in A$, $\mathcal{M}[G^S_w] \models \mathrm{D}_H\,\varphi \leftrightarrow \mathrm{B}_i\,\mathrm{D}_H\,\varphi$.

□

Thus, the only thing that the operation might change concerns the agents' beliefs about the beliefs of the sharing agents.

4.4 Sharing and resolving

In the two-step belief resolution this section discusses, the belief G-sharing update is the first step. Once such an update is performed, cautious belief G-resolution becomes 'safe' in belief models. Indeed, after the G-sharing operation, the beliefs of all agents in G become, in a sense, common knowledge. Thus, when agents in G try to merge their beliefs, the only potential source of conflicts are factual (i.e., first-order) beliefs as well as beliefs about the beliefs of agents not in G. In such cases, rejecting the conflicting information does not affect 'the properties of belief': seriality, transitivity and Euclidicity, are all preserved.

Proposition 4.6. *Take $\varnothing \subset G \subseteq A$. If $\mathcal{M}$ is a belief model, so is $\mathcal{M}[G^S_w][G^{\forall}]$.*

Proof. See the appendix. □

The previous result shows that belief sharing followed by cautious resolution gives a belief-preserving two-step update. We can now look at how this idea works

Example 4.3. *As a first look at the "sharing-then-cautiously-resolving" approach at work, recall the following intuition: an agent a who is uncertain about p, mistakenly believes b believes that p is true and mistakenly believes she is consistent with b, should intuitively believe that, after resolving, she will believe p. As discussed on* *Page 21* *(concerning the pointed model $(\mathcal{M}_2, w)$, below on the left), this fails for the patched cautious resolution.*

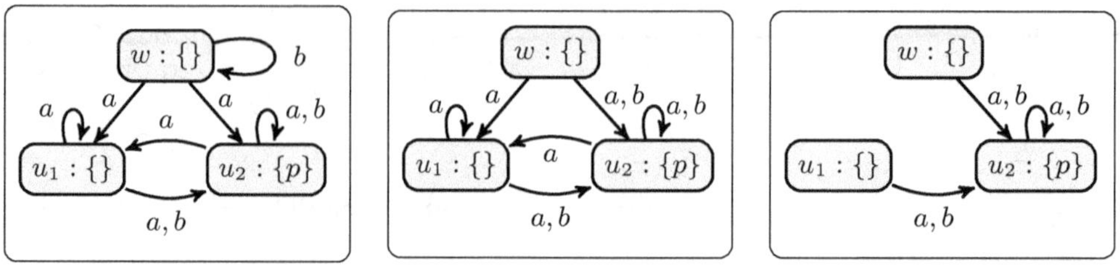

$\mathcal{M}_2$ $\quad$ $\mathcal{M}_2[\{a,b\}^S_{u_{1/2}}]$ $\quad$ $\mathcal{M}_2[\{a,b\}^S_{u_{1/2}}][\{a,b\}^{\forall}]$

Consider now the formula expressing the analogous idea in the two-steps approach, namely $\mathrm{B}_a\, \mathbf{S}_{\{a,b\}}[\{a,b\}!^{\forall}]\, \mathrm{B}_a\, p$. *To decide its truth-value at* $(\mathcal{M}_2, w)$, *one needs to evaluate the subformula* $\mathbf{S}_{\{a,b\}}[\{a,b\}!^{\forall}]\, \mathrm{B}_a\, p$ *at both* $(\mathcal{M}_2, u_1)$ *and* $(\mathcal{M}_2, u_2)$, *which in turn requires to evaluate* $[\{a,b\}!^{\forall}]\, \mathrm{B}_a\, p$ *at both* $(\mathcal{M}_2[\{a,b\}^S_{u_1}], u_1)$ *and* $(\mathcal{M}_2[\{a,b\}^S_{u_2}], u_2)$. *As it turns out, the models* $\mathcal{M}_2[\{a,b\}^S_{u_1}]$ *and* $\mathcal{M}_2[\{a,b\}^S_{u_2}]$ *are the same (above center). Moreover, from there, a further cautious update produces the model above on the right. Since, in the latter,* $\mathrm{B}_a\, p$ *holds at both* u_1 *and* u_2, *we have then that* $[\{a,b\}!^{\forall}]\, \mathrm{B}_a\, p$ *holds at* w *at both* $(\mathcal{M}_2[\{a,b\}^S_{u_1}], u_1)$ *and* $(\mathcal{M}_2[\{a,b\}^S_{u_2}], u_2)$. *Thus,* $\mathrm{B}_a\, \mathbf{S}_{\{a,b\}}[\{a,b\}!^{\forall}]\, \mathrm{B}_a\, p$ *holds at* $(\mathcal{M}_2, w)$.

◀

There is, however, one issue that stems from the relational representation of distributed information. As mentioned in Section 2, distinct worlds may satisfy exactly the same formulas. Thus, disjointness at the level of conjecture sets need not coincide with inconsistency at the level of beliefs. This phenomenon, a well-known limitation of relational models for representing distributed information (again, see, e.g., 18, 36, 41), can be induced by the G-sharing update, which changes accessibility without looking at the worlds' contents. By doing so, it might produce models in which cautious resolution does not combine information in the expected way.

Example 4.4. *Consider the pointed model* $(\mathcal{M}_2, w)$ *below on the left. In it, while agent* a *is uncertain about* p *and believes* b *believes* p *(i.e.,* $\mathcal{M}_2, w \models \neg\, \mathrm{B}_a\, p \wedge \neg\, \mathrm{B}_a\, \neg p \wedge \mathrm{B}_a\, \mathrm{B}_b\, p$*), agent* b *believes both* $\neg p$ *and that agent* a *is uncertain about* p *(i.e.,* $\mathcal{M}_2, w \models \mathrm{B}_b\, \neg p \wedge \mathrm{B}_b(\neg\, \mathrm{B}_a\, p \wedge \neg\, \mathrm{B}_a\, \neg p)$*).*

$\mathcal{M}_2$

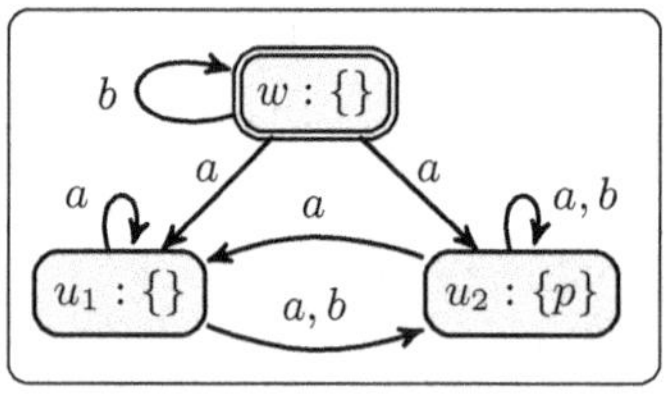

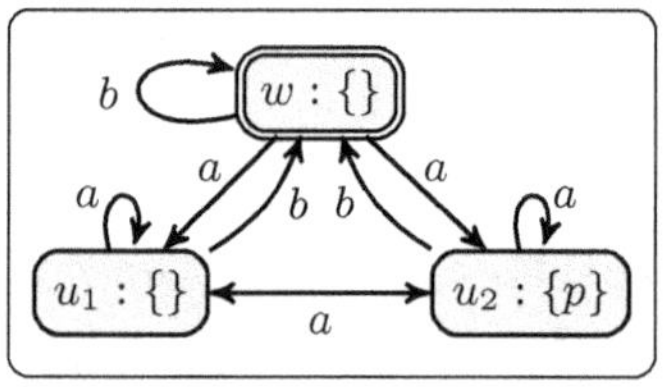

$\mathcal{M}_2[\{a,b\}^S_w]$

After both agents share their beliefs (above on the right), there is no formula on which a and b have opposite views (their 'syntactic' beliefs are consistent). Still, their conjecture sets are disjoint, and thus the cautious G-resolution will not change anything.

However, consider the model $\mathcal{M}_7$ (below center). It is the (individual, i.e., standard) bisimulation contraction *of $\mathcal{M}_2[\{a, b\}^S_w]$ and, moreover, it is (individually) bisimilar to it. Crucially, in $\mathcal{M}_7$, the disjointness of conjecture sets that does not have corresponding inconsistency at the level of individual beliefs (i.e., the individual bisimilarity of w and u_1) has been removed.*[16] *This is a model in which the agents can consistently combine information, yielding the model $\mathcal{M}_7[\{a, b\}^\forall]$ (below right) in which a not only accepts that b believes $\neg p$, but also resolves her uncertainty about p, coming to believe $\neg p$ (i.e., $\mathcal{M}_7[\{a, b\}^\forall], w \vDash \mathrm{B}_a \mathrm{B}_b \neg p \wedge \mathrm{B}_a \neg p$).*

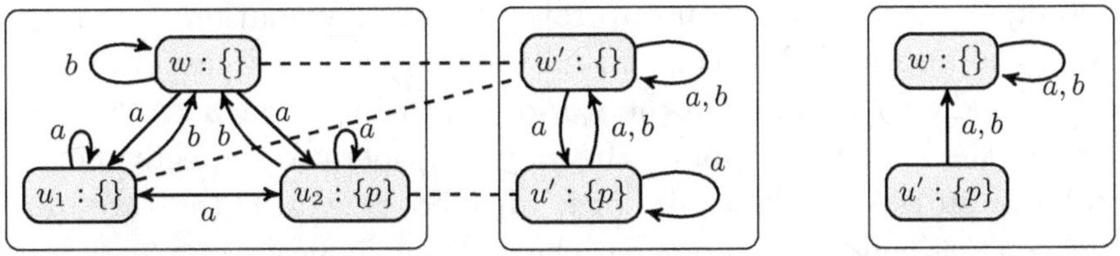

$\mathcal{M}_2[\{a, b\}^S_w]$ $\mathcal{M}_7$ $\mathcal{M}_7[\{a, b\}^\forall]$

◀

Applying cautious resolution after bisimulation contraction yields the expected outcome. Thus, the two-step approach—belief sharing followed by cautious resolution—behaves here as intended once indistinguishable states are identified.

5 Conclusions & further work

This paper concerns the action of information communication. While other papers have discussed variants of this idea in which the discussed information is knowledge (understood here as truthful information), the focus here is the case in which the discussed information are beliefs. We started by showing that the techniques used in the knowledge case have limitations when working with belief (serial, transitive and Euclidean) models, especially when the agents' beliefs are in conflict: if a consistent agent believes a given φ while another believes $\neg\varphi$, the plain combination of such information states will lead to inconsistencies (semantically, the model operation does not preserve seriality). To address this, the paper introduced and analyzed

[16]See, e.g., Gerbrandy [18], Roelofsen [36], van der Hoek et al. [11], for classes of relational models in which different worlds provide different information.

three belief-update operations: cautious belief resolution, patched cautious belief resolution, and belief sharing.

The first proposed operation, cautious belief resolution, works intuitively by combining an agent's information only with those from agents with whom they are collectively consistent. In this way, two agents with opposite opinions can simply ignore each other. With this, consistent agents remain consistent (semantically, the operation preserves seriality). The price to pay here is that, by rejecting information from conflicting agents, the agent might lose introspection properties (semantically, the operation preserves neither transitivity nor Euclidicity). The second proposed operation, patched cautious belief resolution, adds a step that 'recovers' the relevant relational properties in a technically natural way. The price to pay here is that, by doing so, agents might end up rejecting all information they receive. More importantly, the approach has conceptual issues: while individual and distributed beliefs are defined only in terms of the worlds that can be reached (by the respective relation) from the evaluation point, the way the operation affects the worlds that are reachable from the evaluation point is affected by worlds that are unreachable.

A brief analysis made the following clear. First, consistent agents might be, as a group, inconsistent. Thus, to preserve consistency when communicating, agents must reject some of the received information. However, while it is reasonable to reject *some* of the information, it is not reasonable to reject *any/all* of it. In particular, while an agent might reject other agents' beliefs about *non-epistemic facts*, she should not reject other agents' beliefs *about those agents' own beliefs*. After all, even if two agents disagree on the truth-value of a given p (and thus, when exchanging information, they might end up rejecting each other's opinion), the communication should allow both of them to correctly identify the beliefs of the other. With this in mind, the belief sharing operation was introduced as a way to correct higher-order beliefs *prior* to cautious resolution, thus enabling 'safe' belief merging without loss of rationality. While the earlier resolution operations build on well-known relational constructions from the dynamic epistemic logic literature, the belief sharing update constitutes a more novel and less explored strategy.

Several questions remain open regarding the operations introduced in this paper. For the resolution operations, the main challenge is to obtain a complete axiomatization of the patched cautious resolution modality. As discussed, this might require an extension of the base 'static' language $\mathcal{L}_{\mathrm{D}}$, so worlds that will become important after the operation can be identified before the latter takes place. Additionally, connections with non-monotonic reasoning (e.g. [illegible], [illegible]), in particular regarding the selection of admissible belief sets and forms of sceptical reasoning, deserve closer scrutiny.

Concerning the belief sharing operation, a number of avenues merit further investigation. To begin with, the expressivity of the language that adds its associated modality to $\mathcal{L}_{\mathrm{D}}$ has yet to be characterized. Moreover, developing a complete axiomatization—especially in interaction with other dynamic modalities, such as that for cautious belief resolution—remains a difficult and important technical task. Still, there are other aspects that need to be addressed. As the examples suggest, the operation might be better suited for models in which distinct but $\mathcal{L}_{\mathrm{D}}$-indistinguishable worlds do not occur, as the finite distinguishing models of van der Hoek et al. [44], the distinguishing and full models of Gerbrandy [18], and the full communication models of Roelofsen [36].[17] If any of these or other similar constraints are imposed on the initial structures, then one would also need model operations under which the chosen conditions are preserved. While one might 'simply' check whether the operations proposed here satisfy this requirement, one could also look for versions of these operations that take advantage of their lack of representational redundancies. Finally, exploring the connections between belief sharing and other update mechanisms—such as public announcement logic or iterated belief revision—could deepen our understanding of rational belief change in multi-agent systems.

References

[1] Thomas Ågotnes and Yì N. Wáng. Resolving distributed knowledge. *Artificial Intelligence*, 252:1–21, 2017. doi: 10.1016/j.artint.2017.07.002.

[2] Thomas Ågotnes and Yì N. Wáng. Group belief. *Journal of Logic and Computation*, 31(8):1959–1978, 2021. doi: 10.1093/logcom/exaa068.

[3] Thomas Ågotnes and Yì N. Wáng. Somebody knows. In Meghyn Bienvenu, Gerhard Lakemeyer, and Esra Erdem, editors, *Proceedings of the 18th International Conference on Principles of Knowledge Representation and Reasoning, KR 2021, Online event, November 3-12, 2021*, pages 2–11, 2021. doi: 10.24963/kr.2021/1.

[4] Grigoris Antoniou and Kewen Wang. Default logic. In Gabbay and Woods [15], pages 517–555.

[5] Philippe Balbiani, Hans van Ditmarsch, Andreas Herzig, and Tiago de Lima. Some truths are best left unsaid. In Thomas Bolander, Torben Braüner, Silvio Ghilardi, and Larry Moss, editors, *Advances in Modal Logic (AiML 2012), Copenhagen, 22/08/12–25/08/12*, volume 9, pages 36–54. College Publications, 2012. URL http://www.irit.fr/~Andreas.Herzig/P/Aiml12.html.

[17] For example, note how, while $\mathcal{M}_2[\{a,b\}^S_w]$ is not a full communication model, both $\mathcal{M}_2$ and $\mathcal{M}_7$ are.

[6] Alexandru Baltag. What is del good for? Workshop on Logic, Rationality and Intelligent Interaction, 2010. URL http://ai.stanford.edu/~epacuit/lograt/esslli2010-slides/copenhagenesslli.pdf.

[7] Alexandru Baltag and Sonja Smets. Learning what others know. In Elvira Albert and Laura Kovács, editors, *LPAR 2020*, volume 73 of *EPiC Series in Computing*, pages 90–119. EasyChair, 2020. doi: 10.29007/plm4.

[8] Alexandru Baltag, Lawrence S. Moss, and Slawomir Solecki. The logic of public announcements, common knowledge, and private suspicions. In *Proceedings of the 7th Conference on Theoretical Aspects of Rationality and Knowledge*, TARK '98, page 43–56, San Francisco, CA, USA, 1998. Morgan Kaufmann Publishers Inc. doi: 10.5555/645876.671885.

[9] Patrick Blackburn, Maarten de Rijke, and Yde Venema. *Modal Logic.* Cambridge University Press, 2002.

[10] Felix Brandt, Vincent Conitzer, Ulle Endriss, Jérôme Lang, and Ariel D. Procaccia, editors. *Handbook of Computational Social Choice.* Cambridge University Press, USA, 2016. doi: 10.5555/3033138.

[11] Phan Minh Dung. On the acceptability of arguments and its fundamental role in nonmonotonic reasoning, logic programming and n-person games. *Artificial Intelligence*, 77(2):321–358, 1995. doi: 10.1016/0004-3702(94)00041-X.

[12] Ronald Fagin, Joseph Y. Halpern, and Moshe Y. Vardi. What can machines know? On the properties of knowledge in distributed systems. *Journal of the ACM*, 39(2): 328–376, 1992. doi: 10.1145/128749.150945.

[13] Ronald Fagin, Joseph Y. Halpern, Yoram Moses, and Moshe Y. Vardi. *Reasoning about knowledge.* The MIT Press, Cambridge, Mass., 1995.

[14] Raul Fervari and Fernando R. Velázquez-Quesada. Introspection as an action in relational models. *Journal of Logical and Algebraic Methods in Programming*, 108:1–23, 2019. doi: 10.1016/j.jlamp.2019.06.005.

[15] Dov M. Gabbay and John Woods, editors. *The Many Valued and Nonmonotonic Turn in Logic*, volume 8 of *Handbook of the History of Logic.* Elsevier, 2007.

[16] Rustam Galimullin and Louwe B. Kuijer. Varieties of distributed knowledge. In Agata Ciabattoni, David Gabelaia, and Igor Sedlár, editors, *Advances in Modal Logic, AiML 2024, Prague, Czech Republic, August 19-23, 2024*, pages 379–400. College Publications, 2024.

[17] Rustam Galimullin and Fernando R. Velázquez-Quesada. Topic-based communication between agents. *Studia Logica*, 2024. doi: 10.1007/s11225-024-10119-z.

[18] Jelle Gerbrandy. *Bisimulations on Planet Kripke*. Phd thesis, University of Amsterdam, Amsterdam, 1999.

[19] Jelle Gerbrandy and Willem Groeneveld. Reasoning about information change. *Journal of Logic, Language, and Information*, 6(2):147–169, 1997. doi: 10.1023/A:1008222603071.

[20] Davide Grossi and Gabriella Pigozzi. *Judgment Aggregation: A Primer*. Number 27 in Synthesis Lectures on Artificial Intelligence and Machine Learning. Morgan & Claypool Publishers, 2014. doi: 10.2200/S00559ED1V01Y201312AIM027.

[21] Joseph Y. Halpern and Yoram Moses. A guide to the modal logics of knowledge and belief: Preliminary draft. In Aravind K. Joshi, editor, *Proceedings of the 9th International Joint Conference on Artificial Intelligence. Los Angeles, CA, USA, August 1985*, pages 480–490. Morgan Kaufmann, 1985. URL http://ijcai.org/Proceedings/85-1/Papers/094.pdf.

[22] Joseph Y. Halpern and Yoram Moses. Knowledge and common knowledge in a distributed environment. *Journal of the ACM*, 37(3):549–587, 1990. doi: 10.1145/79147.79161.

[23] Andreas Herzig, Emiliano Lorini, Elise Perrotin, Fabián Romero, and François Schwarzentruber. A logic of explicit and implicit distributed belief. In *ECAI 2020 - 24th European Conference on Artificial Intelligence*, volume 325 of *Frontiers in Artificial Intelligence and Applications*, pages 753–760. IOS Press, 2020. doi: 10.3233/FAIA200163.

[24] Risto Hilpinen. Remarks on personal and impersonal knowledge. *Canadian Journal of Philosophy*, 7(1):1–9, 1977. doi: 10.1080/00455091.1977.10716173.

[25] Jaakko Hintikka. *Knowledge and Belief: An Introduction to the Logic of the Two Notions*. Cornell University Press, Ithaca, N.Y., 1962.

[26] Sébastien Konieczny and Ramón Pino Pérez. Logic based merging. *Journal of Philosophical Logic*, 40(2):239–270, March 2011. doi: 10.1007/s10992-011-9175-5.

[27] David Lewis. *Convention: A Philosophical Study*. Harvard University Press, Cambridge, Massachusetts, 1969.

[28] Churn-Jung Liau. A modal logic framework for multi-agent belief fusion. *ACM Transactions on Computational Logic*, 6(1):124–174, 2005. doi: 10.1145/1042038.1042043.

[29] John Lindqvist, Fernando R. Velázquez-Quesada, and Thomas Ågotnes. Towards resolving distributed beliefs. In José Proença, Raul Fervari, Manuel A. Martins, Reinhard Kahle, and Graham Pluck, editors, *Software Engineering and Formal Methods. SEFM 2024 Collocated Workshops*, pages 77–91, Cham, 2025. Springer Nature Switzerland. doi: 10.1007/978-3-031-94748-3_6.

[30] John I. Lindqvist, Fernando R. Velázquez-Quesada, and Thomas Ågotnes. Variations on distributed belief. *Journal of Logic, Language and Information*, 2025. doi: 10.1007/s10849-025-09446-8.

[31] Christian List and Ben Polak. Introduction to judgment aggregation. *Journal of Economic Theory*, 145(2):441–466, 2010. doi: 10.1016/j.jet.2010.02.001.

[32] David Makinson. *Bridges From Classical to Nonmonotonic Logic*. King's College Publications, London, 2005.

[33] Pedrito Maynard-Reid II and Yoav Shoham. Belief fusion: Aggregating pedigreed belief states. *Journal of Logic, Language and Information*, 10(2):183–209, February 2001. doi: 10.1023/A:1008375000210.

[34] Jan A. Plaza. Logics of public communications. In M. L. Emrich, M. S. Pfeifer, M. Hadzikadic, and Z. W. Ras, editors, *Proceedings of the 4th International Symposium on Methodologies for Intelligent Systems*, pages 201–216, Tennessee, USA, 1989. Oak Ridge National Laboratory, ORNL/DSRD-24.

[35] Raymond Reiter. A logic for default reasoning. *Artificial Intelligence*, 13(1-2):81–132, 1980. doi: 10.1016/0004-3702(80)90014-4.

[36] Floris Roelofsen. Distributed knowledge. *Journal of Applied Non-Classical Logics*, 17: 255–273, 01 2007. doi: 10.3166/jancl.17.255-273.

[37] Floris Roelofsen and Yanjing Wang. Contraction semantics for distributed knowledge. Available at https://www.irit.fr/~Andreas.Herzig/Esslli05/FinalPapers/Esslli05Rolofsen.pdf, 2005.

[38] Wolfgang Spohn. Ordinal conditional functions: A dynamic theory of epistemic states. In William Leonard Harper and Brian Skyrms, editors, *Causation in decision, belief change, and statistics: Proceedings of the Irvine Conference on Probability and Causation*, number 42 in The Western Ontario Series in Philosophy of Science, pages 105–134. Kluwer, Dordrecht, 1988. doi: 10.1007/978-94-009-2865-7_6.

[39] Robert Stalnaker. On logics of knowledge and belief. *Philosophical Studies*, 128(1): 169–199, March 2006. doi: 10.1007/s11098-005-4062-y.

[40] David Steiner. A system for consistency preserving belief change. In Sergei Artemov and Rohit Parikh, editors, *Proceedings of Workshop on Rationality and Knowledge. 18th European Summer School of Logic, Language and Information*, pages 133–144, 2006.

[41] Johan van Benthem. *Logical Dynamics of Information and Interaction*. Cambridge University Press, 2011.

[42] Johan van Benthem and Eric Pacuit. Dynamic logics of evidence-based beliefs. *Studia Logica*, 99(1):61–92, 2011. doi: 10.1007/s11225-011-9347-x.

[43] Johan van Benthem, Jan van Eijck, and Barteld Kooi. Logics of communication and change. *Information and Computation*, 204(11):1620–1662, November 2006. doi: 10.1016/j.ic.2006.04.006. URL http://www.illc.uva.nl/Publications/ResearchReports/PP-2005-09.text.pdf.

[44] Wiebe van der Hoek, Bernd van Linder, and John-Jules Meyer. Group knowledge is not always distributed (neither is it always implicit). *Mathematical Social Sciences*, 38 (2):215–240, 1999. doi: 10.1016/S0165-4896(99)00013-X.

[45] Hans van Ditmarsch, Wiebe van der Hoek, and Barteld Kooi. *Dynamic Epistemic Logic*. Springer, Dordrecht, Netherland, 2007.

[46] Fernando R. Velázquez-Quesada. Reliability-based preference dynamics: lexicographic upgrade. *Journal of Logic and Computation*, 27(8):2341–2381, 2017. doi: 10.1093/logcom/exx019.

[47] Fernando R. Velázquez-Quesada. Communication between agents in dynamic epistemic logic. *arXiv CoRR*, abs/2210.04656, 2022. doi: 10.48550/arXiv.2210.04656.

[48] Yanjing Wang and Qinxiang Cao. On axiomatizations of public announcement logic. *Synthese*, 190(Suppl 1):103–134, 2013. doi: 10.1007/s11229-012-0233-5.

A Appendix

A.1 Auxiliary material on closure operations (Subsection 3.2)

Definition A.1 (Closure operations). *Let $R \subseteq W \times W$ be a binary relation.*

- *The relation $R^+ \subseteq W \times W$, called R's* transitive closure, *is defined as*

$$R^+ := \bigcup_{n \geq 1} R^n \qquad \textit{with} \begin{cases} R^1 := R \\ R^{k+1} := R^k \,;\, R \end{cases}$$

  for ";" the standard relational composition.

- *The relation $R^E \subseteq W \times W$, called R's* Euclidean closure, *is defined as*

$$R^E := R \cup (\overleftarrow{R} \,;\, (R \cup \overleftarrow{R})^* \,;\, R) \qquad \textit{with} \begin{cases} R^* := \{(w,w) \mid w \in W\} \cup R^+ \\ \overleftarrow{R} := \{(v,w) \mid Rwv\} \end{cases}$$

Given a model $\mathcal{M}$, use $\mathcal{M}^+$ and $\mathcal{M}^E$ to denote, respectively, the models resulting from replacing each R_i in $\mathcal{M}$ with R_i^+ and R_i^E. ◀

The relation R^+/R^E is, indeed, the smallest transitive/Euclidean relation containing R (for the latter see, e.g., Fervari and Velázquez-Quesada , Lemma 3.1).

There are important aspects to consider before using these operations. First, both closures are needed. On the one hand, transitive closure alone is not enough: there are models (e.g., $\mathcal{M}_2[\{a,b\}^{\forall}]$ in Example 3.1) that are transitive (so, transitive closure will have no effect) but not Euclidean. On the other hand, Euclidean closure alone is not enough: there are models (e.g., the Euclidean closure of the model $\mathcal{M}_3[\{a,b,c\}^{\forall}]$ from Example 3.1, shown below) that are Euclidean (so, Euclidean closure will have no effect) but not transitive.

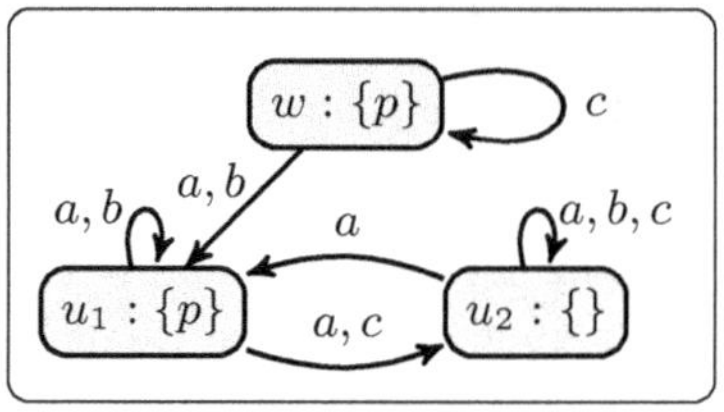

$(\mathcal{M}_3[\{a,b,c\}^{\forall}])^E$

Second, the order in which the closure operations are applied matters for the preservation of both properties, as Euclidean closure does not preserve transitivity. The model $\mathcal{M}_8$ (below on the left) is transitive but not Euclidean (e.g., Rwu and Rww hold, yet Ruw fails); yet, while its Euclidean closure (below on the right) is Euclidean, it is not transitive ($R^E vu$ and $R^E uw$ hold, yet $R^E vw$ fails).

$\mathcal{M}_8$ 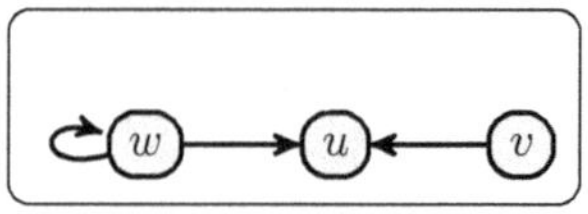

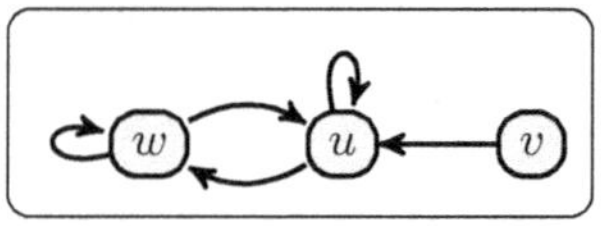

$(\mathcal{M}_8)^E$

Still, as it is shown below, transitive closure preserves Euclidicity.

Proposition A.1. *Let $R \subseteq W \times W$ be a binary relation. If R is Euclidean, so is R^+.*

Proof. Assume R is Euclidean. Take any $w, u_1, u_2 \in W$ with R^+wu_1 and R^+wu_2. By the construction of R^+, while the first implies there are $n \geq 1$ and $\{x_i \mid 1 \leq i \leq n\}$ such that Rwx_1, Rx_ix_{i+1} for $i \in [1 \mathinner{..} n-1]$ and $x_n = u_1$, the second implies there are $m \geq 1$ and $\{y_i \mid 1 \leq i \leq m\}$ such that Rwy_1, Ry_iy_{i+1} for $i \in [1 \mathinner{..} m-1]$ and $y_m = u_2$. It will be shown that Ru_1u_2, which implies the required $R^+u_1u_2$.

Without loss of generality, take $n \leq m$. Note how Rx_iy_i and Ry_ix_i hold for every $i \in [1 \mathinner{..} n]$; this is proved by induction. For the base case ($i = 1$), by R's Euclidicity, the given Rwx_1 and Rwy_1 imply Rx_1y_1 and Ry_1x_1. For the inductive case, suppose

Rx_iy_i and Ry_ix_i for $i \leq n-1$. From the latter, the x-chain yields Rx_ix_{i+1}; from the same, the assumption $n \leq m$ implies $i \leq m-1$, so the y-chain implies Ry_iy_{i+1}. Now, use R's Euclidicity: Rx_iy_i and Rx_ix_{i+1} imply Ry_ix_{i+1}, which together with Ry_iy_{i+1} implies both $Rx_{i+1}y_{i+1}$ and $Ry_{i+1}x_{i+1}$, concluding this induction. This can be pushed forward: Rx_ny_i and Ry_ix_n hold for every $i \in [n..m]$. This is also proved by induction. The base case (Rx_ny_n and Ry_nx_n) is the previous induction's last stage. For the inductive case, suppose Rx_ny_i and Ry_ix_n for $i \leq m-1$. From the latter, the y-chain yields Ry_iy_{i+1} which, together with Ry_ix_n implies (by R's Euclidicity) both $Ry_{i+1}x_n$ and Rx_ny_{i+1}, finishing this induction. Its final stage contains Rx_ny_m, that is, Ru_1u_2, which completes the proof. □

Thus, if transitive closure is applied after Euclidean closure, the final model is guaranteed to be both transitive (by transitive closure) and Euclidean (by Euclidean closure and the preservation result above).

A.2 Omitted proofs

Proof of Proposition 3.2

The validity of Item *(1)* follows from the fact that cautious belief resolution does not affect atomic valuations. That of Items *(2)* and *(3)* is a consequence of the operation being a total function; that of Item *(5)* is straightforward, as indistinguishability for agents not in G remains the same.

For the validity of Item *(4)*, take an arbitrary pointed model $\mathcal{M}, w$ with $\mathcal{M} = \langle W, \{R_i \mid i \in A\}, V\rangle$, and take $G, H \subseteq A$ such that $G \cap H \neq \varnothing$.

(⇒) Suppose $\mathcal{M}, w \vDash [G!^{\forall}]\,\mathrm{D}_H\,\varphi$. Then, $\mathcal{M}[G^{\forall}], w \vDash \mathrm{D}_H\,\varphi$ and thus every world $u \in W$ which satisfies $R_i[G^{\forall}]wu$ for every $i \in H$ is such that $\mathcal{M}[G^{\forall}], u \vDash \varphi$. Since $G \cap H \neq \varnothing$, we can partition H into $H \cap G$ and $H \setminus G$. Thus, following the definition of $R_i[G^{\forall}]$, every world $u \in W$ which satisfies

(i) $u \in R_{G'}(w)$ for some G' such that $i \in G' \subseteq_w^{max} G$ for every $i \in H \cap G$, and
(ii) $u \in R_i(w)$ for every $i \in H \setminus G$

is such that $\mathcal{M}[G^{\forall}], u \vDash \varphi$. Call this the main left-to-right assumption.

Now, take any $F_1 \subseteq G \cap H$ and $F_2 \subseteq G$ such that $\mathcal{M}, w \vDash max_G(F_1 \cup F_2)$. Then, $F_1 \cup F_2 \subseteq_w^{max} G$. Further, take any $u \in W$ such that $u \in R_{F_2 \cup H}(w)$; it will be shown that u satisfies $\mathcal{M}, u \vDash [G!^{\forall}]\varphi$, so $\mathcal{M}, w \vDash \mathrm{D}_{F_2 \cup H}[G!^{\forall}]\varphi$, as required. For this, we require to parts. ***(i)*** First, note how $u \in R_{F_2 \cup H}(w)$ implies both $u \in R_{F_2 \cup \{i\}}(w)$ for every $i \in H$ as well as $u \in R_H(w)$. From the former, every $i \in H$ is consistent with F_2 at w, and thus so is every $i \in H \cap G$. From the latter, every $i \in H \cap G$ is consistent with F_1 at w, as the assumed $F_1 \subseteq G \cap H$ implies $F_1 \subseteq H$. Hence, every $i \in H \cap G$ is consistent with $F_1 \cup F_2$ at w. But $F_1 \cup F_2$ is already maximally consistent (w.r.t. G) at w; thus, $H \cap G \subseteq F_1 \cup F_2$. In other words, for every $i \in H \cap G$ there is $F_1 \cup F_2$ which is maximally consistent (w.r.t. G) at w and which satisfies $i \in F_1 \cup F_2$; crucially, $u \in R_{F_1 \cup F_2}(w)$ as, recall, $u \in R_{F_2 \cup H}(w)$ (for $u \in R_{F_2}(w)$) as well as $u \in R_H(w)$ and $F_1 \subseteq H$ (for $u \in R_{F_1}(w)$). ***(ii)*** Second, take any $i \in H \setminus G$. Since $i \in H$, from $u \in R_{F_2 \cup H}(w)$ it follows that $u \in R_{\{i\}}(w)$. From these two pieces and the main left-to-right assumption above, it follows that $\mathcal{M}[G^{\forall}], u \vDash \varphi$.

($\Leftarrow$) Suppose $\mathcal{M}, w \vDash \bigwedge_{F_1 \subseteq G \cap H} \bigwedge_{F_2 \subseteq G} (max_G(F_1 \cup F_2) \to \mathrm{D}_{F_2 \cup H}[G!^{\forall}]\varphi)$. To show that $\mathcal{M}, w \vDash [G!^{\forall}]\,\mathrm{D}_H\,\varphi$ holds, it will be shown that $\mathcal{M}[G^{\forall}], w \vDash \mathrm{D}_H\,\varphi$, that is, that every $u \in R_H[G^{\forall}](w)$ is such that $\mathcal{M}[G^{\forall}], u \vDash \varphi$. So, take any $u \in R_H[G^{\forall}](w)$, that is, take any $u \in W$ satisfying $R_i[G^{\forall}]wu$ for every $i \in H$. Since $G \cap H \neq \varnothing$, we can partition H into $H \cap G$ and $H \setminus G$. Thus, following the definition of $R_i[G^{\forall}]$, the given u satisfies Item *(i)* and Item *(ii)* above (it is still left to prove that it satisfies $\mathcal{M}[G^{\forall}], u \vDash \varphi$). From Item *(i)*, every $i \in H \cap G$ has an $G' \subseteq^{max}_{w} G$ with $i \in G'$ s.t. $u \in R_{G'}(w)$. For each such i, its respective $G' \subseteq^{max}_{w} G$ can be partitioned into its H-part $F_1 = H \cap G$ (so $i \in F_1$) and its non-H part $F_2 = G' \setminus F_1$; thus, for each such i there are $F_1 \subseteq H \cap G$ and $F_2 \subseteq G$ such that $F_1 \cup F_2 \subseteq^{max}_{w} G$ with $i \in F_1 \cup F_2$ s.t. $u \in R_{F_1 \cup F_2}(w)$. Thus, in particular, $u \in R_{F_2}(w)$. Moreover, note how $u \in R_i(w)$ for all $i \in H$: indeed, if $i \in H \cap G$ then this follows from Item *(i)*, and if $i \in H \setminus G$ then this follows from Item *(ii)*. Thus, $u \in R_H(w)$, with together with the previous $u \in R_{F_2}(w)$ implies $u \in R_{F_2 \cup H}(w)$. But then, by the initial assumption for this right-to-left direction, $\mathcal{M}[G^{\forall}], u \vDash \varphi$. This concludes the proof.

Proof of Proposition 3.4

Take $i \notin G$. Then, $R_i[G^{\forall}] = R_i$, and thus $R_i[G^{\forall}]$ is Euclidean (because R_i is). Then, $(R_i[G^{\forall}])^E = R_i[G^{\forall}]$, and thus $(R_i[G^{\forall}])^E$ is transitive (because $R_i[G^{\forall}] = R_i$ is). Hence $((R_i[G^{\forall}])^E)^+ = (R_i[G^{\forall}])^E$. Therefore, $R_i[G^{\odot}] = R_i$ so $R_i[G^{\odot}] \subseteq R_i$.

Take $i \in G$. Then, $R_i[G^{\forall}] \subseteq R_i$.[18] In the rest, use $S := R_i$, $S' := R_i[G^{\forall}]$, $S'^E := (R_i[G^{\forall}])^E$ and $S'^{E+} := ((R_i[G^{\forall}])^E)^+$.

- First, it will be shown that $S'^E \subseteq S$. For this, and using $S' \subseteq S'^E$ (as Euclidean closure does not remove edges), partition S'^E into S' and $S'^E \setminus S'$. For the former, any pair in S' is also in S (the cautious update does not add edges). For the latter, take any $(u_1, u_2) \in S'^E \setminus S'$. By definition, $(u_1, u_2) \in (\overleftarrow{S'}\ ;\ (S' \cup \overleftarrow{S'})^*\ ;\ S')$. Then, there are $n \geq 0$ and $\{x_k \mid 0 \leq k \leq n\}$ such that $S'x_0u_1$, $S'x_nu_2$, and $S'x_kx_{k+1}$ or $S'x_{k+1}x_k$ for $k \in [0\,..\,n-1]$. Since $S' \subseteq S$, the $n \geq 0$ and $\{x_k \mid 0 \leq k \leq n\}$ are

[18] If $R_i[G^{\forall}]wu$, then $R_H wu$ for some $H \subseteq^{max}_{w} G$ with $i \in H$. From the latter and R_H's definition ($\bigcap_{b \in H} R_b$), $R_H wu$ implies $R_i wu$.

such that Sx_0u_1, Sx_nu_2, and Sx_kx_{k+1} or $Sx_{k+1}x_k$ for $k \in [0 .. n-1]$. Now, note how Sx_ku_1 for every $k \in [0 .. n]$.[19] Hence, Sx_nu_1, which together with Sx_nu_2 imply the required Su_1u_2, as S is Euclidean. Therefore, $S'^E \subseteq S$.

- Now, it will be shown that $S'^{E+} \subseteq S$. For this, and using $S'^E \subseteq S'^{E+}$ (as transitive closure does not remove edges), partition S'^{E+} into S'^E and $S'^{E+} \setminus S'^E$. For the former, any pair in S'^E is also in S, as it has just been shown above. For the latter, take any $(u, u') \in S'^{E+} \setminus S'^E$. By definition, there are $m \geq 1$ and $\{y_k \mid 1 \leq k \leq m\}$ such that $S'^E uy_1$, $S'^E y_ky_{k+1}$ for $k \in [1 .. m-1]$, and $y_m = u'$. Since $S'^E \subseteq S$, the $m \geq 1$ and $\{y_k \mid 1 \leq k \leq m\}$ are such that Suy_1, Sy_ky_{k+1} for $k \in [1 .. m-1]$, and $y_m = u'$. This yields the required Suu', as S is transitive. Therefore, $S'^{E+} \subseteq S$.

Thus, $((R_i[G^{\forall}])^E)^+ \subseteq R_i$.

Proof of Proposition 4.6

Take any serial, transitive and Euclidean pointed model $(\mathcal{M}, w)$ with $\mathcal{M} = \langle W, \{R_i \mid i \in A\}, V\rangle$.

- Take any $i \notin G$. Both operations leave the relations R_i untouched, so trivially any property of R_i will hold also of $R_i[G^S_w][G^{\forall}]$.

- Take any $i \in G$. Since R_i is serial, $R_i[G^S_w]$ is serial (Proposition 4.1), and thus $\mathcal{M}[G^S_w][G^{\forall}]$ is serial (Proposition 3.1). For transitivity and Euclidicity, we rely on an intermediate result. Recall Proposition 4.3: $R_H[G^S_w](v_1) = R_H[G^S_w](v_2)$ for all $H \subseteq G$ and all $v_1, v_2 \in W$. Thus, in $\mathcal{M}[G^S_w]$, for all H such that $i \in H$, we have $H \subseteq^{max}_{v_1} G$ if and only if $H \subseteq^{max}_{v_2} G$, and $R_H[G^S_w](v_1) = R_H[G^S_w](v_2)$. Then, for all $v_1, v_2 \in W$, $R_i[G^S_w][G^{\forall}](v_1) = R_i[G^S_w][G^{\forall}](v_2)$. Now, to see that $R_i[G^S_w][G^{\forall}]$ is transitive, take arbitrary $x, y, z \in W$ with $y \in R_i[G^S_w][G^{\forall}](x)$ and $z \in R_i[G^S_w][G^{\forall}](y)$. From $R_i[G^S_w][G^{\forall}](x) = R_i[G^S_w][G^{\forall}](y)$, the required $z \in R_i[G^S_w][G^{\forall}](x)$ follows. Then, to see that it is Euclidean, take arbitrary $x, y, z \in W$ with $y \in R_i[G^S_w][G^{\forall}](x)$ and $z \in R_i[G^S_w][G^{\forall}](x)$. From $R_i[G^S_w][G^{\forall}](y) = R_i[G^S_w][G^{\forall}](x)$, the required $z \in R_i[G^S_w][G^{\forall}](y)$ follows.

[19] The base case of this is the given Sx_0u_1. For the inductive case, suppose Sx_ku_1 for $k < n$ (so x_{k+1} exists). If Sx_kx_{k+1} then $Sx_{k+1}u_1$ as S is Euclidean; if $Sx_{k+1}x_k$ then $Sx_{k+1}u_1$ as S is transitive.

Received June 2021

Paraconsistent Reactive Graphs: Their Semantics and Conflict Resolution Policies

Juliana Cunha
CIDMA, Dep. Mathematics, Aveiro University, Portugal
INESC TEC & Dep. Informatics, Minho University, Portugal
juliana.cunha@ua.pt

Alexandre Madeira
CIDMA, Dep. Mathematics, Aveiro University, Aveiro, Portugal
madeira@ua.pt

Luís Soares Barbosa
INESC TEC & Dep. Informatics, Minho University, Portugal
lsb@di.uminho.pt

Abstract

This paper introduces *Paraconsistent Reactive Graphs*, an extension of Reactive Graphs that incorporates paraconsistency into the graph edges to naturally address the vagueness and inconsistency often encountered in dynamic systems. In these systems, the dynamic behavior is given by a special kind of edges, called *hyper-edges*, which act as switches that either activate or deactivate others. Furthermore, edges in Paraconsistent Reactive Graphs are not bivalent; instead, the framework allows for the representation of vague, consistent, and inconsistent information. This is potentially relevant when modeling inputs from multiple sources that may be individually consistent but collectively incomplete or conflicting.

Consequently, this work raises fundamental questions about conflict resolution, namely, how to manage and combine information even in the presence of contradictions. To address this, we propose and motivate, a few strategies for combining information from various sources. Finally, a semantics for Paraconsistent Reactive Graphs is presented and results under which consistency is preserved are explored.

This work is supported by CIDMA (https://ror.org/05pm2mw36) under the Portuguese Foundation for Science and Technology (FCT, https://ror.org/00snfqn58), Grants UID/04106/2025 (https://doi.org/10.54499/UID/04106/2025) and UID/PRR/04106/2025 and with projects COMPETE2030-FEDER-00892000. Cunha is also supported by the FCT PhD Grant 2024.04652.BD.

1 Introduction

Dov Gabbay introduced *reactive graphs* as transition structures whose accessibility relation can be dynamically modified by prior transitions [27, 29]. This dynamic behavior is specified by higher-order edges, or hyper-edges, which update the accessibility relation when traversed. Indeed hyper-edges connect one edge to another, acting as switches that either activate ($\twoheadrightarrow$) or deactivate ($\rightarrow\!\!\times$) the connected edge [26]. Thus, a reactive graph consists of ordinary transitions, called *ground edges*, enriched with hyper-edges that control activation of other edges. Reactive graphs have proven effective in modeling a wide range of reactive systems [27, 28, 41].

However, certain practical applications involve input from multiple, possibly conflicting sources. In such contexts, the classical bivalent structure of reactive graphs may be insufficient. This motivates the exploration of extensions that can handle vague and inconsistent information in a sound and controlled manner.

One such extension introduces fuzziness in both ground and hyper-edges which are assigned values in the real interval $[0,1]$, capturing varying degrees of certainty [7, 8, 9, 41]. This framework allows modeling systems with vague behavior that can be explored using a fuzzy modal logic. Furthermore, this was applied to model scenarios in biology [41], for example, the circadian rhythm of cyanobacteria, a biological system represented by a regulatory network of components such as genes and proteins. In such networks, one component may induce (via activation) or inhibit (via deactivation) the production of another. Moreover, transitions in these systems may occur with varying degrees of likelihood, making fuzzy reactive graphs a natural and expressive framework for their specification.

Building on this idea, we propose a paraconsistent extension of reactive graphs that accommodates both uncertainty and inconsistency. In this work, the term *paraconsistency* is used in a precise sense to indicate that the framework tolerates contradictions without leading to trivialization. Formally, paraconsistency is a property of a consequence relation: a logic is said to be paraconsistent if its consequence relation is *non-explosive*, meaning that from a contradiction $A, \neg A$, it is not possible to derive an arbitrary conclusion B. Thus, paraconsistent reasoning offers a sound and controlled way to reason in the presence of contradictions.

One of the founding contributions to the field of paraconsistency is Jaśkowski's *discursive logic*, introduced in 1948 [?]. Jaśkowski was inspired by group discussions in which participants may each provide self-consistent information, yet when combined may yield contradictions. Since Jaśkowski's work, extensive developments by Asenjo, Carnielli, da Costa, and others have established paraconsistent reasoning as a well-developed area of study [3, 10, 20], see also [25] for a comprehensive overview.

Our starting point is Belnap-Dunn four-valued lattice [5], where each truth value is a pair $(\mathit{tt}, \mathit{ff}) \in \{0,1\}^2$. Informally, as discussed in [1], such a pair can be understood so that tt represents the amount of information in favour of a statement and ff the amount of information against it. Hence, the truth value $(1,0)$ stands for "true", in the sense that there is information only in favour of the statement; $(0,1)$ stands for "false", information only against it; $(1,1)$ stands for both ("contradiction"); and $(0,0)$ stands for neither ("vagueness"). These values can be ordered by increasing truth $\leqslant_t$ and decreasing knowledge $\leqslant_k$ as shown in the diagrams below:

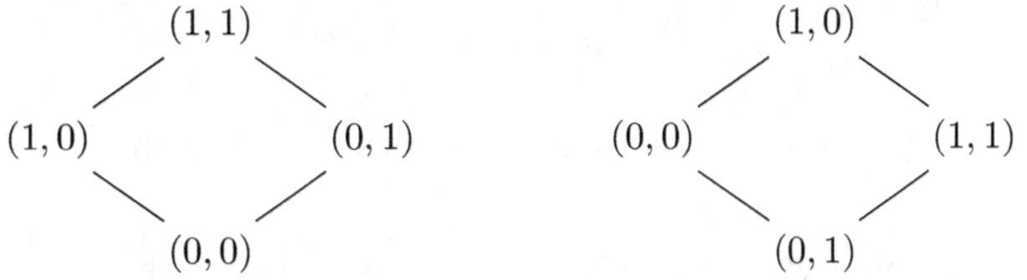

Lattice $\mathcal{FOUR}$ with $\leqslant_k$-order   Lattice $\mathcal{FOUR}$ with $\leqslant_t$-order

The framework underlying this paper generalizes $\mathcal{FOUR}$ by allowing weights to be parameterized over a class of lattices. Given a complete lattice $(A, \leqslant)$, one can construct a complete lattice A^2 as the product of A with its order dual A^∂. This construction, known as a *twist structure*, a term coined by Kracht [37], has been extensively studied in the literature [31, 35, 37]. In particular, Fitting [23] showed that it provides a general method for constructing bilattices, i.e., structures $(B, \leqslant_t, \leqslant_k, \neg)$ equipped with two orders $\leqslant_t$ and $\leqslant_k$, such that $(B, \leqslant_t)$ and $(B, \leqslant_k)$ are complete lattices, and $\neg$ is a unary involutive operator on B that is increasing w.r.t. $\leqslant_t$ and decreasing w.r.t. $\leqslant_k$ [1]. The notion of a logical bilattice and the corresponding (paraconsistent) logic system was introduced by Arieli and Avron [1] with applications to preference modelling [22].

Our line of research [15, 16, 18] focuses on *square bilattices* [22], where values are interpreted as pairs $(\mathit{tt}, \mathit{ff}) \in A^2$, where tt represents the amount of belief on an assertion and ff the degree of belief against it. For example, suppose that weights for both transitions come from a complete lattice over the interval $[0,1]$, see Figure 1. Such pairs of weights capture different behaviors:

- **Inconsistency**, when the positive and negative weights are contradictory, i.e., they sum to a value greater than 1, which corresponds to the upper triangle in Figure 1.
- **Vagueness**, when the sum is less than 1, corresponding to the lower triangle in Figure 1.

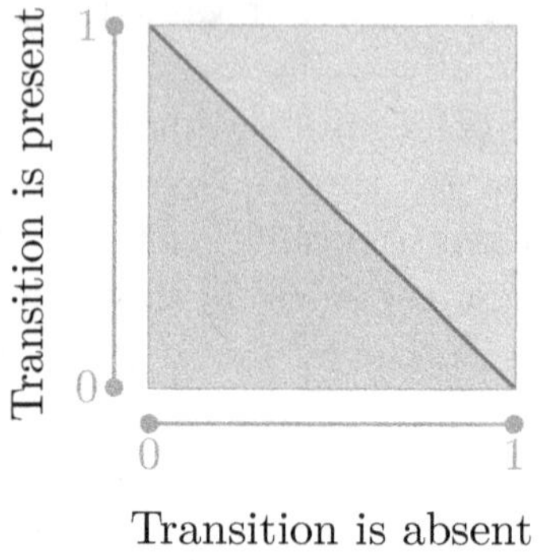

Figure 1: The vagueness-inconsistency square

- **Consistency**, when the sum is exactly 1, which means that the belief for and against an assertion are complementary, corresponding to the anti-diagonal line in Figure 1.

While fuzzy approaches are well-suited to represent vague information, our paraconsistent framework additionally accommodates inconsistencies. For instance, as discussed in research on bilattices applied to preference modeling [22], let $P(u, v)$ be a relation that expresses whether u is (known to be) at least as good as v. In a fuzzy setting, assigning $P(u, v) = 0$ may either mean that u is definitely not preferred to v, or that there is simply no information available to make a comparison. This ambiguity motivates the use of a pair of values rather than a single fuzzy value, in order to distinguish more precisely between different informational scenarios: *truth*, *falsity*, *ignorance*, and *contradiction*.

In many contexts, contradictions arise naturally from conflicting evidence, misinformation or evolving beliefs, and often go hand in hand with dynamic behavior. A legal trial is a classic example, frequently used in research in discursive logic, see [40, Section 4.2.]. In such settings, multiple witnesses may offer individually consistent testimonies that collectively conflict. Expert witnesses, due to their specialized knowledge, may refute earlier claims, thereby influencing which conclusions modeled as transitions (i.e., edges in the model) are considered reliable. Over time, trust in different sources may shift. For instance, if conflicting testimonies are given about a weapon and an expert rules out some possibilities, their input should affect the system by activating or deactivating certain transitions. Moreover, a witness credibility is not fixed: if a statement is discredited, future claims from that source may carry less weight. These evolving states highlight the natural connection between paraconsistency and dynamic behavior. A suitable modeling framework must therefore accommodate contradictions while adapting as new information becomes available.

Prior work [11] introduced paraconsistency into reactive graphs by employing Belnap's four-valued propositional variables. In addition, they proposed a paraconsistent logic and explored related topics, such as the measurement of inconsistencies, building on earlier contributions [12, 13]. In contrast, our approach incorporates paraconsistency directly into the edges of the graph, rather than into the propositional variables. This is achieved by allowing each edge to be weighted by a pair of values $(tt, ff) \in 2^2$. It is convenient to think of the pair (tt, ff) as a pair of extreme truth values, where tt represents the truth value of the statement "the transition will occur," and ff represents the truth value of the statement "the transition will be prevented from occurring." As a result, this framework can naturally represent vague, consistent, or inconsistent information, as described above. Consequently, all the relevant constructions in our framework are parametric in a class of lattices $(A, \leqslant)$. This accommodates different instantiations depending on the set A of possible values best suited to model the problem at hand.

This paper begins by introducing the necessary background definitions required for the sequel, including those related to Reactive Graphs, paraconsistent graphs as introduced in [15], and *twist-structures*, which are essential for reasoning with pairs of values. As introduced in [17], we then combined the latter graph notions to define the concept of a *Reactive Graph with Paraconsistent Transitions* (abbreviated as $\mathrm{RG_P}$). An $\mathrm{RG_P}$ is a graph in which ground transitions are weighted by pairs of truth values representing evidence for and against the occurrence of a transition, and where bivalent hyper-edges act as switches that activate or deactivate other edges. Additionaly in the original paper that this work extends [17], motivated by the work in [42], we presented a semantics for $\mathrm{RG_P}$ that supports the "unfolding" [30] of an $\mathrm{RG_P}$ into a paraconsistent graph, and provided a concrete application of the model. Building on this earlier conference paper [17], we extend this framework to introduce paraconsistent reactive graphs (abbreviated as PRG), in which both ground and hyper-edges are weighted by pairs of values. This generalization reflects the idea that the influence of hyper-edges may itself be uncertain or contradictory, rather than strictly bivalent.

In this paper, we begin by studying PRG under the assumption that hyper-edges are non-conflicting, i.e., no two hyper-edges are triggered by the same edge and act upon the same target. This restriction allows us to understand more precisely the impact that activating and deactivating hyper-edges have on the system's behavior. This extension requires a careful revision of the definitions originally presented in [17].

In a subsequent section, we address scenarios in which conflicting hyper-edges may arise. Since edges may encode information from diverse sources, it is natural to encounter situations where some sources support the activation of a transition while others support its deactivation. This raises important questions concerning conflict management within paraconsistent graphs. Moreover, throughout the paper, we explore results concerning consistency preservation under the proposed semantics.

In conclusion, we extend the work presented in the original conference paper [17] with the following main contributions:

1. We define paraconsistent reactive graphs, in which both ground and hyper-edges can represent vague, consistent or inconsistent information, extending the reactive graphs with paraconsistent transitions in [17], in which only the ground edges can represent non-bivalent information.

2. We develop a semantics for paraconsistent reactive graphs, first without conflicting hyper-edges, and then for the general case.

3. We prove monotonicity and consistency-preserving results for both semantics, with the latter following as a consequence of the novel Theorems 1 and 2.

4. We propose a number of conflict resolution policies, each supported by a clear motivation.

The remainder of the paper is organized as follows: Section 2 presents background definitions on reactive graphs, paraconsistent graphs and twist constructions which are essential for the sequel. Section 3 recalls the definition of Reactive Graphs with Paraconsistent transitions and their semantics, as introduced in the original conference paper [17]. Section 4 introduces paraconsistent reactive graphs and their semantics. The later section is divided in two: Section 4.1 outlines the formalism and results whenever paraconsistent reactive graphs have no conflicting hyper-edges and Section 4.2 addresses conflicts. Finally, Section 5 offers concluding remarks and identifies several lines for future research.

2 Preliminaries

We begin this section by recalling the definition of a *reactive graph*, as presented in [42]. We then present twist-structures as a construction parameterized by a class of lattices $(A, \leqslant)$ that enables the computation of pairs of values in A^2. Next, we recall the definition of a paraconsistent graph, introduced in [15]. Finally, in order to give a concrete meaning to the vagueness-inconsistency square depicted in Figure 1 we endow the underlying lattice $(A, \leqslant)$ with a metric d, [14, 18].

The main goal of this section is to lay the groundwork for the subsequent sections, which aim to combine reactive graphs (see Definition 1) and paraconsistent graphs (see Definition 4) into a unique structure that captures the dynamic behavior of the former and the representation of inconsistencies in the latter.

2.1 Reactive Graphs

In this section, we recall the definition of a reactive graph as introduced in [42], along with its operational semantics.

Reactive graphs were proposed in [42] as an extension of reactive graphs originally introduced by Gabbay [27]. A reactive graph is a graph in which transitions between states, called ground edges ($\twoheadrightarrow$), occur via the execution of an action. In addition to ground edges, these graphs are also endowed with higher-order edges, known as hyper-edges, that represent the system's reactive behavior. Hyper-edges connect one edge to another and act as switches: whenever an edge is traversed, it may trigger a hyper-edge that either activates ($\twoheadrightarrow\!\!\!\!\rightarrow$) or deactivates ($\rightarrow\!\!\!\!\times$) another.

Definition 1. *Let Σ be a set of action symbols. A Σ-reactive graph is a tuple $M = (W, w_0, E, \mathcal{C}_0, \twoheadrightarrow, \twoheadrightarrow\!\!\!\!\rightarrow, \rightarrow\!\!\!\!\times, \bar{\cdot})$ where*

- *W is the set of states and $w_0 \in W$ is the initial state;*
- *E is the set of edge names and $\mathcal{C}_0 \subseteq E$ is the set of initially active edges;*
- *$\twoheadrightarrow \subseteq W \times \Sigma \times W$ is the set of ground edges; $\twoheadrightarrow\!\!\!\!\rightarrow \subseteq E \times E$ is the set of activating edges; $\rightarrow\!\!\!\!\times \subseteq E \times E$ is the set of deactivating edges;*
- *$\bar{\cdot} : E \longrightarrow (\twoheadrightarrow \dot{\cup} \twoheadrightarrow\!\!\!\!\rightarrow \dot{\cup} \rightarrow\!\!\!\!\times)$ is a bijective function that maps edges in E to their definition.*

Graphically, edges that are not active are represented by a dashed arrow.

The following example illustrates how a vending machine can be modeled by a Σ-reactive graphs as defined above.

Example 1 ([42]). *The following reactive graph models the behavior of a buyer interacting with a vending machine that offers two products: coffee and chocolate. The buyer has only 1 euro to spend and can either purchase two coffees (at 0.5€ each) or one chocolate (at 1€).*

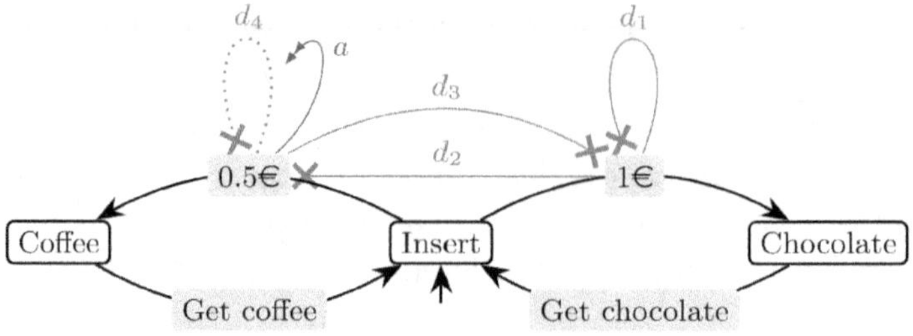

Intuitively, in the initial state, if the buyer performs the action 1€, *he immediately spends all the available funds and buys a chocolate. This triggers the action of deactivating edges* d_1 *and* d_2*, which disable any further action from the initial state. As a result, although the system returns to the initial state after the buyer gets a chocolate, no additional actions are available. Graphically, this is reflected by the fact that only dashed arrows, corresponding to inactive edges, originate from the initial state.*

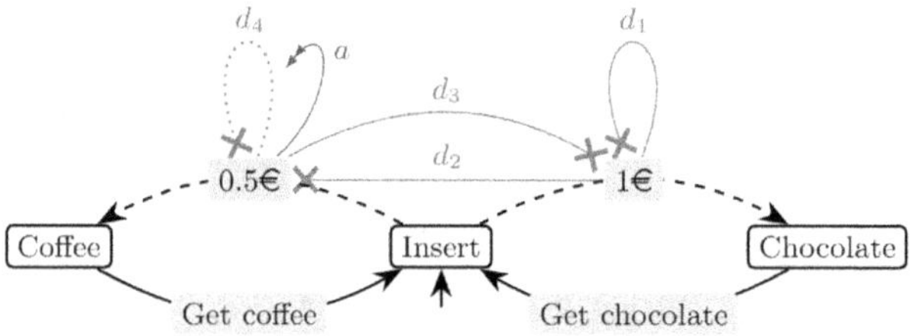

Alternatively, if the buyer first performs the action 0.5€, *this triggers the deactivation edge* d_3*, which disables the edge labeled* 1€, *preventing the buyer from purchasing a chocolate since there are insufficient funds. This action also triggers the activation edge* a*, which activates edge* d_4*, allowing the buyer to perform only one more* 0.5€ *action, thus enabling the purchase of a second coffee. Graphically, after executing the action* 0.5€ *once, the system appears as follows:*

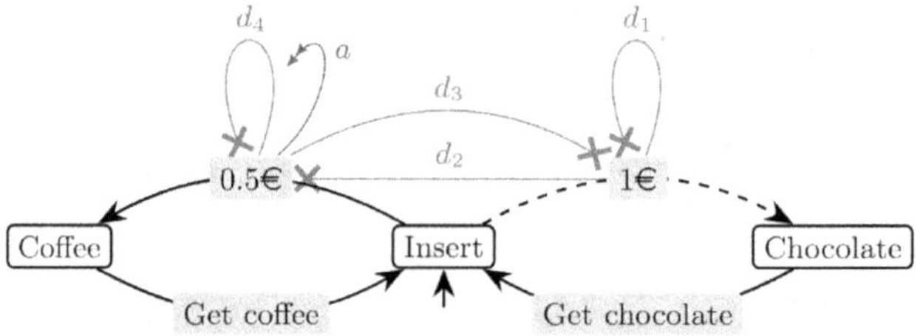

The buyer decision to purchase a second coffee, triggers edge d_4*, which disables any further coffee purchases, as the buyer has run out of funds. Graphically:*

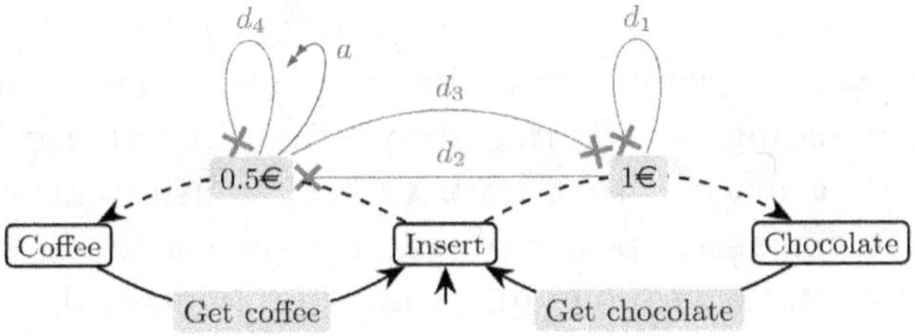

Formally, we define this as a Σ-reactive graph $M = (W, w_0, E, \mathcal{C}_0, \rightarrow, \twoheadrightarrow, \multimap, \bar{\cdot})$, *where* $\Sigma = \{1€, 0.5€, \textit{Get coffee}, \textit{Get chocolate}\}$ *is the set of action symbols;* $W = \{\textit{Insert}, \textit{Coffee}, \textit{Chocolate}\}$ *is the set of states; and* $w_0 = \textit{Insert}$ *is the initial state. The set of edge names is* $E = \{1€, 0.5€, \textit{Get coffee}, \textit{Get chocolate}, d_1, d_2, d_3, d_4, a\}$. *We slightly abuse notation by naming the ground edges after their corresponding action symbols to improve readability.* $\mathcal{C}_0 = E \backslash \{d_4\}$ *denotes the set of initially active edges. The set of ground edges* $\rightarrow$ *are those depicted in black in the graphs; the activating edges* $\twoheadrightarrow$ *are depicted in blue; and the deactivating edges* $\multimap$ *are shown in red. Finally, the function* $\bar{\cdot}$ *defines the correspondence between edge names and their internal behavior. For example, the ground edge named* 0.5€ *is mapped to* $\langle \textit{Insert}, 0.5€, \textit{Coffee} \rangle$, *indicating that this ground edge goes from state Insert to Coffee by the action* 0.5€. *While the deactivation edge* d_2 *is mapped to its source and target, that is,* $\overline{d_2} = \langle 1€, 0.5€ \rangle$, *and similarly, the activation edge* a *is mapped as* $\overline{a} = \langle 0.5€, d_4 \rangle$.

Next we recall a few concepts originally introduced in [42], which play a crucial role in defining the operational semantics of reactive graphs.

Given a reactive graph M, an edge $e_s \in E$, and the current set of active edges $\mathcal{C} \subseteq E$. Then, $\mathsf{from}(e_s)$ denotes the set of hyper-edges that directly originate from e_s. Additionally, $\mathsf{from}^*(e_s, \mathcal{C})$ denotes the set that recursively transverses from to collect all active hyper-edges,i.e., edges in $\mathcal{C}$, triggered by e_s.

$$\begin{aligned} \mathsf{from}(e_s) &= \{e \mid \exists e_t \cdot \overline{e} = (e_s, e_t)\} \\ \mathsf{from}^*(e_s, \mathcal{C}) &= \bigcup_{r \in (\mathsf{from}(e_s) \cap \mathcal{C})} \mathsf{from}^*(r, \mathcal{C} \backslash \{e_s\}) \cup \{r\} \end{aligned} \tag{1}$$

where $\mathsf{from}^{\star}(r, \emptyset) = \emptyset$ for any $r \in E$.

Additionally, the set of edges activated by some edge e (respectively deactivated by e), written $\mathsf{on}(e, \mathcal{C})$ (respectively $\mathsf{off}(e, \mathcal{C})$) is defined as follows [42].

$$\begin{aligned} \mathsf{on}(e, \mathcal{C}) &= \{e_t \mid e_{trg} \in \mathsf{from}^*(e, \mathcal{C}) \wedge \exists e_s \cdot \overline{e_{trg}} = (e_s, e_t) \in \twoheadrightarrow\} \\ \mathsf{off}(e, \mathcal{C}) &= \{e_t \mid e_{trg} \in \mathsf{from}^*(e, \mathcal{C}) \wedge \exists e_s \cdot \overline{e_{trg}} = (e_s, e_t) \in \multimap\} \end{aligned} \tag{2}$$

Given a reactive graph, it is possible to unfold its reactive behavior into an ordinary graph—that is, a graph with no hyper-edges. In fact, Gabbay's work introduced not only an unfolding method that transforms a reactive graph into a standard one, but also a folding method that reverses this transformation. Together, these techniques show the expressive power of reactive graphs: folding can exponentially reduce the number of states of the graph, while unfolding enables the application of standard techniques, such as those used in the axiomatization of logics over reactive frames [29].

In this paper, we focus exclusively on the unfolding process that transforms a reactive graph into a standard graph. Before introducing this transformation through the operational semantics presented in Definition 2, we first formally define what it means for a reactive graph to evolve from one configuration to another.

For a Σ-reactive graph $M = (W, w_0, E, \mathcal{C}_0, \rightarrow, \twoheadrightarrow, \rightarrow\!\!\times, \bar{\cdot})$, a *configuration* is a tuple $\langle w, \mathcal{C}\rangle$, where $w \in W$ is a state and $\mathcal{C} \subseteq E$ is a set of active edges.

Given the inherently dynamic behavior of reactive graphs, in which traversing some ground edges may activate or deactivate other's, the system can evolve from an initial configuration $\langle w_0, \mathcal{C}_0\rangle$ to a new one $\langle w, \mathcal{C}\rangle$. We refer to each such transition, from one configuration to another, as an evolution. Intuitively, an evolution records how the system changes after a single traversal of a ground edge. Formally, given two configurations $\langle w, \mathcal{C}\rangle$ and $\langle w', \mathcal{C}'\rangle$, we say that $\langle w, \mathcal{C}\rangle$ evolves to $\langle w', \mathcal{C}'\rangle$ if there exists an active ground edge connecting w to w', and the set of active edges is updated from $\mathcal{C}$ to $\mathcal{C}'$ after traversing that edge. This notion provides the foundation for the unfolding process, which transforms a reactive graph into a standard one by systematically considering all possible evolutions starting from the initial configuration.

Definition 2 ([42]). *The semantics of a graph M results from the evolution of the initial configuration $\langle w_0, \mathcal{C}_0\rangle$, given by the rule below.*

$$\frac{\exists e \in \mathcal{C} \;\cdot\; \overline{e} = (w, a, w') \;\wedge\; \mathcal{C}' = \big(\mathcal{C} \cup \mathsf{on}(e, \mathcal{C})\big)\backslash\mathsf{off}(e, \mathcal{C})}{\langle w, \mathcal{C}\rangle \xrightarrow{a}_M \langle w', \mathcal{C}'\rangle}$$

where $\langle w, \mathcal{C}\rangle$ and $\langle w', \mathcal{C}'\rangle$ are states of M and there is a transition in M from $\langle w, \mathcal{C}\rangle$ to $\langle w', \mathcal{C}'\rangle$ triggered by action a.

Example 2 ([42]). *Using the semantics defined in Definition 2 to the reactive graph in Example 1 we obtain the following graph.*

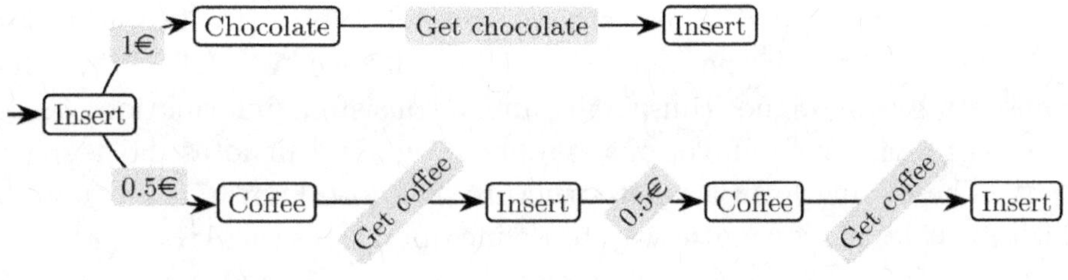

Note that each state in the above standard graph is of the form $\langle w, C\rangle$, where C is the set of active edges updated by the traversal of an edge. However, to improve readability, configurations are omitted from the state labels.

2.2 Paraconsistent Graphs

Before introducing the notion of a paraconsistent graph parameterized by a class of lattices, let us review the underlying semantic structures.

As in recent works on many-valued frameworks [6], our approach is parameterized over a class of complete lattices. Each lattice $\mathbf{A} = (A, \leqslant)$ acts over a non-empty set A of possible values, equipped with binary operations $\sqcap$ (meet) and $\sqcup$ (join) satisfying the usual distributive laws.

Example 3. *Well-known examples of lattices used throughout this paper include*[1]*:*

- *The two element Boolean algebra* $\mathbf{2} = (\{0,1\}, \leqslant_{\mathbf{2}})$ *where* $0 \leqslant_{\mathbf{2}} 1$, *and the meet and join operations are standard Boolean conjunction and disjunction.*
- *The Łukasiewicz three-valued algebra* $\mathbf{3} = (\{1,u,0\}, \leqslant_{\mathbf{3}})$ *where* $0 \leqslant_{\mathbf{3}} u \leqslant_{\mathbf{3}} 1$. *The intermediate value u stands for "unknown" and is also denoted in the literature by $1/2$ or #, often interpreted as "possibility" or "indeterminacy" [40]. The meet and join operations are defined as follows:*

$\sqcap_{\mathbf{3}}$	0	u	1
0	0	0	0
u	0	u	u
1	0	u	1

$\sqcup_{\mathbf{3}}$	0	u	1
0	0	u	1
u	u	u	1
1	1	1	1

- *The Gödel algebra* $\mathbf{G} = ([0,1], \leqslant_{\mathbf{G}})$ *where $\sqcap$ and $\sqcup$ correspond to the minimum and maximum functions on $[0,1]$, respectively. For any $x, y \in [0,1]$, $x \leqslant_{\mathbf{G}} y$ if and only if $x = \min(x,y)$.*

[1] Indices are omitted when clear from context.

As discussed above, this framework was extended in in [15, 18] by working with pairs of values $(tt, ff) \in A^2$, parameterized over a lattice $\mathbf{A} = (A, \leqslant)$, which enables the representation of vague, consistent, and inconsistent information. Here, tt denotes the degree of belief in favor of a statement, while ff denotes the degree of belief against it. These pairs form a twist-structure [23], also known as a square bilattice [2], which provides a systematic way to define operations on A^2.

Definition 3. *Let $\mathbf{A} = (A, \leqslant)$ be a complete distributive lattice over a non-empty set of values A. The structure $\mathcal{A} = (A \times A, \leqslant_t, \leqslant_k, \neg)$ is $\mathbf{A}$-twist structure where $\neg(tt, ff) = (ff, tt)$ and*

- $(tt_1, ff_1) \leqslant_t (tt_2, ff_2)$ *if and only if* $tt_1 \leqslant tt_2$ *and* $ff_1 \geqslant ff_2$
- $(tt_1, ff_1) \leqslant_k (tt_2, ff_2)$ *if and only if* $tt_1 \leqslant tt_2$ *and* $ff_1 \leqslant ff_2$

Denoting the join and meet operations of $\mathbf{A}$ by $\sqcap$ and $\sqcup$ respectively, the following operators are defined for any pair $(tt_1, ff_1), (tt_2, ff_2) \in A^2$:

$$
\begin{aligned}
(tt_1, ff_1) \mathbin{\overline{\sqcap}} (tt_2, ff_2) &= (tt_1 \sqcap tt_2, ff_1 \sqcup ff_2) \\
(tt_1, ff_1) \mathbin{\underline{\sqcup}} (tt_2, ff_2) &= (tt_1 \sqcup tt_2, ff_1 \sqcap ff_2) \\
(tt_1, ff_1) \otimes (tt_2, ff_2) &= (tt_1 \sqcap tt_2, ff_1 \sqcap ff_2) \\
(tt_1, ff_1) \oplus (tt_2, ff_2) &= (tt_1 \sqcup tt_2, ff_1 \sqcup ff_2)
\end{aligned}
$$

The order relation $\leqslant$ in $\mathbf{A}$ induces two orders on A^2: the truth order $\leqslant_t$ and the knowledge order $\leqslant_k$. Operations $\overline{\sqcap}$ and $\underline{\sqcup}$ correspond to the meet and join under $\leqslant_t$, intuitively representing conjunction ("and") and disjunction ("or"), respectively. Operations $\otimes$ and $\oplus$ correspond to meet and join under $\leqslant_k$ and have distinct epistemic interpretations [1]. Specifically, $\otimes$ represents consensus, so that $p \otimes q$ yields the highest amount of information on which both p and q agree with respect to the knowledge order $\leqslant_k$, while $\oplus$ can be viewed as guillibity or "credulousness", accepting the combined knowledge provided by p and q. For instance, $(1, 0) \otimes (0, 1) = (0, 0)$, since one value represents complete belief in a statement, $(1, 0)$, and the other complete disbelief, $(0, 1)$; so the most they can agree on is nothing, $(0, 0)$. In contrast, $(1, 0) \oplus (0, 1) = (1, 1)$, as the combination of both values represents simultaneous belief and disbelief, resulting in a contradiction. Finally, the unary operator $\neg$ corresponds to an involutive negation [1]. Intuitively, if the evidence for and against a statement is tt and ff, respectively, then the evidence for and against its negation is ff and tt, respectively. As expected, this negation reverses truth; that is, if $(tt, ff) \leqslant_t (tt', ff')$, then $\neg(tt, ff) \geqslant_t \neg(tt', ff')$. Moreover, it preserves knowledge: if $(tt, ff) \leqslant_k (tt', ff')$, then $\neg(tt, ff) \leqslant_k \neg(tt', ff')$, so that one knows neither more nor less about a statement than about its negation [24]. These operators play a central role in the Section 4, to combine information from multiple edges in paraconsistent graphs.

Since $\mathbf{A} = (A, \leqslant)$ is distributive, so is the corresponding twist-structure $\mathcal{A} = (A \times A, \leqslant_t, \leqslant_k, \neg)$ [22, Proposition 2]. That is, all twelve distributive laws involving $\sqcap$, $\sqcup$, $\otimes$, and $\oplus$ hold. For example, for any values (tt_1, ff_1), (tt_2, ff_2) and (tt_3, ff_3) in A^2, the following distributive law holds:

$$((tt_1, ff_1) \sqcap (tt_2, ff_2)) \otimes (tt_3, ff_3) = ((tt_1, ff_1) \otimes (tt_3, ff_3)) \sqcap ((tt_2, ff_2) \otimes (tt_3, ff_3))$$

As previously stated, this framework underpins our earlier work [15, 18], where we introduced the notion of a paraconsistent graph, i.e., a graph in which transitions are labeled by pairs of values in A^2. Many-valued reasoning is often parametric on lattices (see, e.g., [6, 33, 38]) or bilattices (see, e.g., [1, 24]), which makes it convenient to refer to their elements as truth values. Accordingly, the pairs $(tt, ff) \in A^2$ can be viewed as representing a space of truth values, intuitively capturing the extent to which the statement "the transition is possible" holds or not, respectively. Let us recall the formal definition of a paraconsistent graph.

Definition 4. *Let $(A, \leqslant)$ be a complete distributive lattice and Σ a set of action symbols. An (A, Σ)-paraconsistent graph is a tuple $M = (W, w_0, R)$ where*

- *W is a non-empty set of states and $w_0 \in W$ is the initial state;*
- *$R \subseteq W \times \Sigma \times A^2 \times W$ is a paraconsistent relation, i.e., if $(w, a, tt, ff, w') \in R$, then tt measures the evidence of the transition from w to w' occurring via action a and ff measures the evidence of the transition being prevented from occurring.*

Example 4. *The structure $(\{w_0, w_1, w_3\}, w_0, R)$ depicted below is a $([0,1], \{a, b\})$-paraconsistent graph, whose transitions are labeled by pairs of truth values in $[0,1]^2$, capturing the extent to which each transition may occur or be prevented from occurring.*

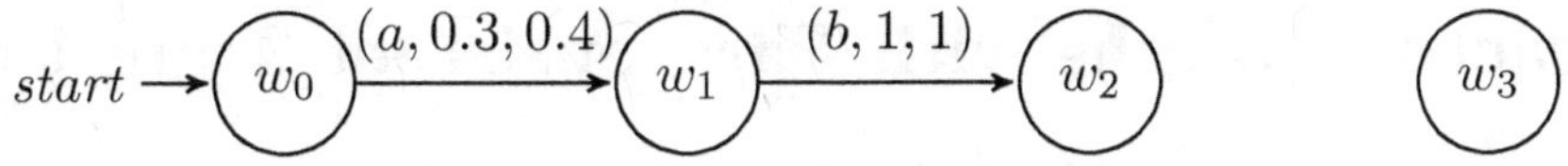

Transition $(w_0, a, 0.3, 0.4, w_1)$ represents vague information, since the sum of weights is less than one, $0.3 + 0.4 < 1$. In contrast, $(w_1, b, 1, 1, w_2)$ represents inconsistent information, as the sum of the weights exceeds 1. Transitions that are known to be stricly false, such as $(w_2, a, 0, 1, w_3)$, are omitted from the diagram for clarity.

Finally, as pointed out in the Introduction, in order to precisely describe the two triangular regions shown in Figure 1 and the diagonal separating them, we endow the underlying lattice **A** with a metric $d : A \to \mathbb{R}$.

Example 5. *For each of the lattices in Ex. 3, we provide an suitable choice of d.*

For the Boolean algebra **2**, *the three valued algebra* **3** *and the Gödel algebra* **G** *consider the following metrics, respectively.*

$$d_{\mathbf{2}}(x,y) = \begin{cases} 0 & \textit{if } x = y \\ 1 & \textit{otherwise} \end{cases}$$

$d_{\mathbf{3}}$	0	u	1
0	*0*	*1*	*2*
u	*1*	*0*	*1*
1	*2*	*1*	*0*

$$d_{\mathbf{G}}(x,y) = \sqrt{(x-y)^2}$$

Each metric d can be lifted to a metric $D : A^2 \to \mathbb{R}$ defined over pairs of values, as expected:

$$D((tt_1, ff_1), (tt_2, ff_2)) = \sqrt{d(tt_1, tt_2)^2 + d(ff_1, ff_2)^2} \tag{3}$$

This metric is particularly relevant as it provides a concrete interpretation for the vagueness-inconsistency square depicted in Figure 1. Consequently, it enables the definition of the consistent and paraconsistent sets, introduced in [15], denoted by Δ_C and Δ_P, respectively.

$$\begin{aligned} \Delta_C &= \{(tt, ff) \mid D((tt, ff), (0,0)) \leqslant D((tt, ff), (1,1))\} \\ \Delta_P &= \{(tt, ff) \mid D((tt, ff), (1,1)) < D((tt, ff), (0,0))\} \end{aligned} \tag{4}$$

The set Δ_C encompasses all pairs within the lower triangle and the black anti-diagonal line in Figure 1, while Δ_P includes pairs located within the upper triangle. Therefore, by comparing the distance of any pair of weights to the elements $(0,0)$ and $(1,1)$, it is possible to determine whether it represents consistent or inconsistent information. Moreover, the set $\Delta = \{(tt, ff) \mid D((tt, ff), (0,0)) = D((tt, ff), (1,1))\}$ includes all pairs equidistant from $(0,0)$ and $(1,1)$, representing *strictly* consistent information.

3 Reactive Graphs with Paraconsistent Transitions

This section revisits the definition of reactive graphs with paraconsistent transitions, introduced in the original conference paper that this work extends [17]. In this framework dynamic behavior is modeled through hyper-edges that serve as switches, either activating ($\twoheadrightarrow$) or deactivating ($\rightarrow\!\!\!\times$) other edges. Although such hyper-edges are not themselves traversable, they encapsulate the system's reactivity by indicating how edges change in response to the traversal of a ground edge ($\twoheadrightarrow$). Paraconsistency is incorporated solely in ground edges, which are weighted by pairs of values in A^2 over a lattice $\mathbf{A} = (A, \leqslant)$.

In this first step, our aim is to combine reactive graphs (see Definition 1) and paraconsistent graphs (see Definition 4) into a novel structure that captures the dynamic behavior of the former and the representation of inconsistencies in the latter. In the next section, we will extend this definition to paraconsistent reactive graphs, where all edges ($\rightarrow$, $\twoheadrightarrow$, $\rightarrow\!\!\!\times$) are weighted by a pair of values in A^2. This generalization is motivated by the observation that switches themselves may carry vague or inconsistent information, arising from incomplete knowledge or internal contradictions. This, in turn, influences the system's behavior.

The definitions presented in this section are adapted from recent work on reactive graphs [42]. We define reactive graphs with paraconsistent transitions as an extension of paraconsistent graphs with hyper-edges. Then, their operational semantics is obtained by "unfolding" [29] them back into standard paraconsistent graphs. This process effectively eliminates hyper-edges while preserving the overall behavior. We conclude the section with a worked example.

Definition 5. *Let $(A, \leqslant)$ be a complete distributive lattice and Σ a set of action symbols. An (A, Σ)-reactive graph with paraconsistent transitions, abbreviated to* $\mathrm{RG_P}$*, is a tuple* $M = (W, w_0, C_0, E, \rightarrow, \twoheadrightarrow, \rightarrow\!\!\!\times, \bar{\cdot})$ *where:*

- *W is a set of states and $w_0 \in W$ is the initial state;*
- *E is the set of all edge names and $C_0 \subseteq E$ is the set of initially active edges;*
- *$\rightarrow \subseteq W \times \Sigma \times A^2 \times W$ is the set of ground edges, $\twoheadrightarrow \subseteq E \times E$ is the set of activating edges and $\rightarrow\!\!\!\times \subseteq E \times E$ is the set of deactivating edges;*
- *$\bar{\cdot} : E \longrightarrow (\rightarrow \dot{\cup} \twoheadrightarrow \dot{\cup} \rightarrow\!\!\!\times)$ is a function that maps an edge in E to its definition*

Example 6. *The system depicted below is a* $([0, 1], \{a, b\})$-$\mathrm{RG_P}$*. The set of all edges is $E = \{e_1, e_2, e_3, e_a, e_d\}$, each marked in yellow in the diagram below. The initial set of active edges is given by $C_0 = E \backslash \{e_2\}$. As before, the inactive edge e_2 is depicted as a dashed arrow. The function $\bar{\cdot}$ maps each edge to its internal details. For instance, for the ground edge e_3 and the hyper-edge e_d, we have $\overline{e_3} = (w_2, a, 1, 0, w_3)$ and $\overline{e_d} = (e_a, e_3)$.*

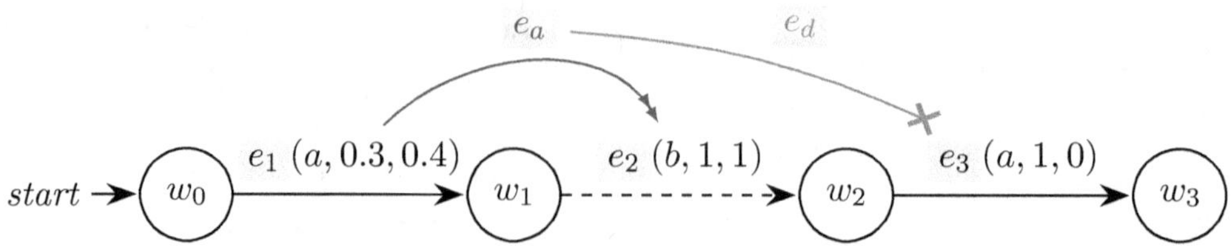

Note that hyper-edges act purely as switches and have no direct impact on the weights of ground transitions.

For an (A, Σ)-RG$_P$, let $\mathcal{C} \subseteq E$ be the current set of active edges, and $e_s \in E$ be some edge. As before, $\mathsf{from}(e_s)$ and $\mathsf{from}^\star(e_s, \mathcal{C})$ denote the sets of hyper-edges triggered by e_s, either directly or recursively, as defined in (1). Additionally, the set of edges activated (resp. deactivated) by e_s, denoted $\mathsf{on}(e_s, \mathcal{C})$ (resp. $\mathsf{off}(e_s, \mathcal{C})$), is defined as in (2). The following example illustrates these definitions in the context of an RG$_P$.

Example 7. *Recall Example 6. Edge e_1 represents a transition from state w_0 to w_1 via action a, with degrees of belief 0.3 for occurrence and 0.4 for non-occurrence. Hence, we have $\overline{e_1} = (w_0, a, 0.3, 0.4, w_1)$. This edge triggers the hyper-edges e_a and e_d, which in turn activate e_2 and deactivate e_3, respectively. We can now compute:*

- $\mathsf{from}(e_1) = \{e_a\}$ *returns the hyper-edges directly triggered by e_1;*
- $\mathsf{from}^*(e_1, \mathcal{C}_0) = \{e_a, e_d\}$ *returns all the hyper edges triggered by e_1 under the initial set of active edges $\mathcal{C}_0$;*
- $\mathsf{on}(e_1, \mathcal{C}_0) = \{e_2\}$ *and* $\mathsf{off}(e_1, \mathcal{C}_0) = \{e_3\}$ *return the edges targeted by activating and deactivating hyper-edges, respectively, triggered by e_1.*

The following definition introduces the operational semantics for the sort of graphs considered in this section. Given an RG$_P$, this semantics unfolds its dynamic behavior into a paraconsistent graph without hyper-edges [29].

Definition 6. *The semantics of an (A, Σ)-paraconsistent graph P, as defined in Definition 4, results from the evolution of the initial configuration $\langle w_0, \mathcal{C}_0 \rangle$, given by the rule below.*

$$\frac{\exists e \in \mathcal{C} \ \cdot \ \overline{e} = (w, a, tt, ff, w') \ \wedge \ \mathcal{C}' = \left(\mathcal{C} \cup \mathsf{on}(e, \mathcal{C})\right) \backslash \mathsf{off}(e, \mathcal{C})}{\langle w, \mathcal{C} \rangle \xrightarrow{(a, tt, ff)}_P \langle w', \mathcal{C}' \rangle}$$

where $\langle w, \mathcal{C} \rangle$ and $\langle w', \mathcal{C}' \rangle$ are states of P connected by a paraconsistent transition via the action symbol a and weighted by the pair (tt, ff).

Applied to Example 6 this yields the standard paraconsistent graph in Example 4.

As in [42], this semantics differs from Gabbay's original one by collecting all activated edges before applying their (de)activation effects, instead of triggering effects immediately during traversal.

Additionally, this semantics introduces a bias: when an edge is both activated and deactivated within the same step, deactivation takes precedence. While this choice simplifies the semantics above, it may not always reflect the intended behavior. We will return to this issue in Section 4.2, when considering paraconsistent reactive graphs with conflicting hyper-edges and propose some policies for managing such conflicts.

The following example illustrates the unfolding of a reactive graph with paraconsistent transitions.

Example 8. *Consider the following* $(\{a,b\},\{0,u,1\} - \mathrm{RG_P}$ *where traversing edge* e_1 *triggers the hyper-edge* e_a *that activates* e_2*, and traversing edge* e_4 *triggers the hyper-edges* e_{d_1} *and* e_{d_2} *that deactivate* e_a *and* e_2*, respectively.*

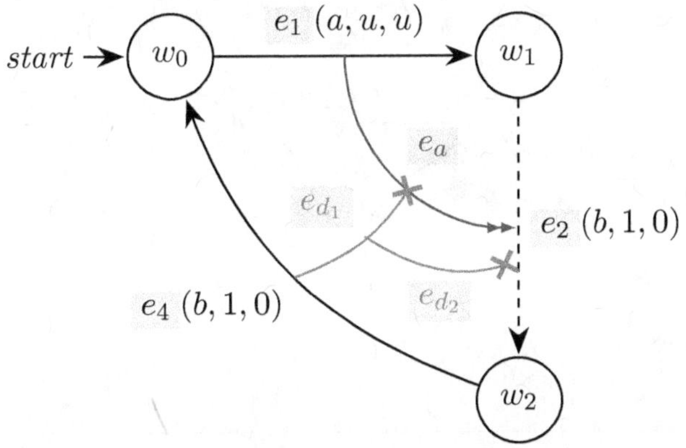

Essentially, we can follow the path $w_0 \xrightarrow{e_1} w_1 \xrightarrow{e_2} w_2 \xrightarrow{e_4} w_0 \xrightarrow{e_1} w_1$*. By the time we return to* w_1*, transition* e_2 *is no longer active, and the hyper-edge that could activate it,* e_a*, has also been deactivated. Hence, there is no other active edge to traverse. More precisely, if we follow the path as described, the initial set of active edges* $C_0 = \{e_1, e_4, e_a, e_{d_1}, e_{d_2}\}$ *is updated to* $C_1 = C_0 \cup \{e_2\}$ *when traversing edge* e_1*, since this edge triggers the activation of* e_2*. Subsequently, traversing edge* e_2 *has no impact on the set of active edges, so the set remains equal to* C_1*. Then, traversing edge* e_4 *updates the active edges from* C_1 *to* $C_2 = C_1 \backslash \{e_a, e_2\}$ *by deactivating* e_a *and* e_2*. Finally, traversing* e_1 *again does not trigger any further activation or deactivation.*

This corresponds exactly to the unfolding of the $\mathrm{RG_P}$ *as defined in Definition 15. Therefore, this* $\mathrm{RG_P}$ *can be unfolded into the following paraconsistent graph.*

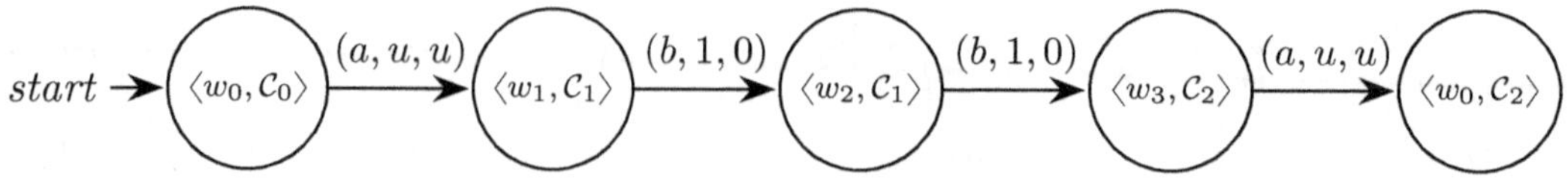

Finally, we present a concrete application of reactive graphs with paraconsistent transitions.

Example 9 (Liar's Coin). *Two players are engaged in a game called* Liar's Coin, *which is adapted from work on epistemic logic [39]. Player 1 bets on heads, while Player 2 bets on tails. Player 1 tosses a coin and announces either heads or tails, but keeps the actual outcome hidden from Player 2. If Player 1 announces tails (†t), Player 2 wins, and the game ends. If Player 1 announces heads (†h), Player 2 has two options: pass (p)– in which case Player 1 wins– or challenge (c), forcing Player 1 to show the coin (h or t). The player with the correct bet then wins. From Player 2's perspective, the game can be represented by the graph[2] in Figure 2.*

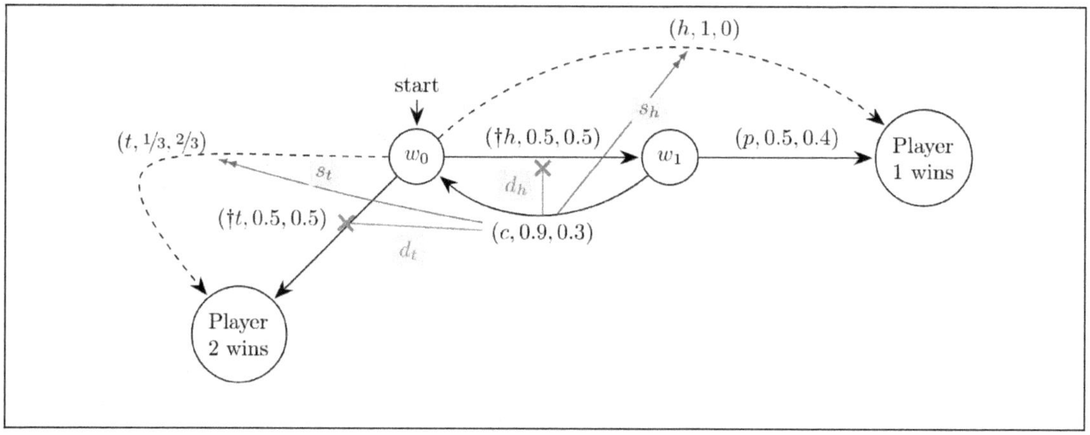

Figure 2

Initially, Player 2's belief about the coin outcome (heads or tails) reflects that of a fair coin toss. Therefore, the transitions representing Player 1 announcing heads or tails, †h and †t, are given by the consistent truth value $(0.5, 0.5)$*, indicating equal uncertainty for and against the outcome.*

When Player 1 announces heads, Player 2 must choose to either challenge or pass. This decision may depend on various vague cues– Player 1's demeanor, prior experience, or intuition. If Player 2 chooses to challenge, the announcement edges,

[2]To improve readability, we omit the names of the ground edges, which are highlighted in yellow. For the original graph, see the conference paper that this work extends [17].

given by the actions $\dagger h$ *and* $\dagger t$*, are deactivated, and the edges given by the actions* h *and* t *are activated. These edges are assigned weights that reflect Player 1's likelihood of honesty. In this example, Player 1 is assumed never to lie about a heads result, so the ground-edge with action* h *has a value of* $(1, 0)$*. For the ground edge with action* t*, however, the evidence is more uncertain: suppose from previous games Player 2 believes Player 1 lies about tails with probability* $1/3$*. Then, revealing tails is weighted with* $(1/3, 2/3)$*.*

Now suppose Player 1 announces "heads". Player 2 must evaluate the evidence for challenging or passing. This evidence may include:

1. *In the last two rounds, Player 1 also announced "heads", which seems unlikely.*
2. *Player 2 knows from prior rounds that Player 1 is generally honest.*
3. *Player 1 appeared nervous during the announcement.*
4. *Player 2 is a risk-taker and inclined to challenge.*

Based on this, Player 2 might subjectively assign non-complementary degrees of belief to each action. These values need not be complementary, as Player 2's reasoning integrates both statistical and intuitive arguments. Using the information from points (1)–(4)*, Player 2 ranks the evidence for and against challenging or passing as follows:*

- *Evidence to challenge is the strongest, say* 0.9*, supported by* (1)*,* (3) *and* (4)*.*
- *Evidence to pass follows, say* 0.5*, moderately supported by* (2) *but tempered by skepticism from* (1)*.*
- *Evidence against passing comes next, say* 0.4*, based partly on* (2) *and* (4)*. While both are vague,* (4) *plays a stronger role in Player 2's behavior.*
- *Evidence against challenging is the weakest, say* 0.3*, primarily based on* (2)*.*

Thus, the edges corresponding to the challenging (c) and passing (p) actions are assigned the values $(0.9, 0.3)$ *and* $(0.5, 0.4)$*, respectively.*

4 Paraconsistent Reactive Graphs

We now extend the definition of reactive graphs with paraconsistent transitions (Definition 5) by allowing hyper-edges ($\twoheadrightarrow$ and $\rightarrow\!\!\!\!\times$) to be also weighted by pairs of truth values in A^2. As a result, all transitions in the graph, including both ground and hyper-edges, may now represent vague, consistent or inconsistent information.

This extension is motivated by scenarios in which information is provided by multiple, possibly unreliable or conflicting, sources. For example, recall the "Liars Coin" example (Example 9). Suppose there are now three players: Player 1 flips the coin and hides the result, while Player 3– whose intentions are unclear– observes the outcome and announces either "heads" or "tails," potentially siding with Player 1 or Player 2 depending on their own interests. In such a case, Player 2 receives contradictory information if Player 1 announces "heads" and Player 3 announces "tails," or vice versa, which can be represented by the pair $(1, 1)$. If Player 2 believes that Player 3's statement undermines Player 1's, they may be incentivized to distrust Player 1 and challenging him– i.e., activating the "challenge" edge. Such dynamics illustrates the need for non-classical hyper-edges that capture uncertainty or inconsistency in the information provided by agents.

This section is organized in two parts:

- Subsection 4.1, introduces paraconsistent reactive graphs (PRG) under the restriction that no conflicting hyper-edges are allowed. Specifically, no two hyper-edges may be triggered by the same source edge and target the same edge. This notion, mathematically formalized, contributes to the understanding of the isolated influence of paraconsistent hyper-edges in a controlled setting.

- Subsection 4.2,addresses the more general and challenging case in which conflicting hyper-edges are permitted. Then, we explore policies for resolving contradictions arising from different sources of information.

Let us now define paraconsistent reactive graphs.

Definition 7. *Let $(A, \leqslant)$ be a complete distributive lattice and Σ a set of action symbols. An (A, Σ)-paraconsistent reactive graph, abbreviated to* PRG, *is a tuple $M = (W, w_0, \rightarrow, \twoheadrightarrow, \multimap, \mathcal{C}_0, \bar{\cdot}, \bar{\bar{\cdot}})$ where:*

- *W is a set of states and $w_0 \in W$ is the initial state;*

- *$\rightarrow$, $\twoheadrightarrow$ and $\multimap$ are the sets of ground, activating, and deactivating edge names, respectively. We define $E = (\rightarrow \cup \twoheadrightarrow \cup \multimap)$ as the set of all edges, and $H = (\twoheadrightarrow \cup \multimap)$ as the set of all hyper-edges.*

- *$\mathcal{C}_0 : E \longrightarrow A^2$ is the initial activation function, assigning to each edge $e \in E$ a pair $(tt, ff) \in A^2$, where tt and ff represent the degrees of belief supporting and opposing the traversal of e, respectively.*

- $\bar{\cdot} : \twoheadrightarrow \longrightarrow (W \times \Sigma \times W)$ *and* $\bar{\bar{\cdot}} : H \longrightarrow E^2$ *are functions such that* $\bar{\cdot}$ *maps each ground edge in* $\twoheadrightarrow$ *to its definition, while* $\bar{\bar{\cdot}}$ *maps each hyper-edge to the pair of edges it connects: the trigger and the target.*

The notation adopted here is slightly different from that used in the previous section for reactive graphs with paraconsistent transitions (see Definition 5). This adjustment is necessary to accommodate the fact that all transitions, both ground edges and hyper-edges, are now weighted by pairs in A^2.

The main change is that the set of initial active edges $\mathcal{C}_0 \subseteq E$ is now replaced by an initial activation function $\mathcal{C}_0 : E \longrightarrow A^2$, which assigns to each edge a pair of values representing the positive and negative evidence for its execution. For example, if $\mathcal{C}_0(e_1) = (0, 1)$, this corresponds to the classical case in which edge e_1 is inactive. Visually, we will no longer use dashed arrows in diagrams to represent inactive edges, as the active/inactive edges are now determined by the values assigned by $\mathcal{C}_0$.

As it will become clear later, it is important to distinguish between ground, activating, and deactivating edges, especially since their values may evolve during execution as the behavior of one edge impacts the weight of others. For this reason, we explicitly define the sets $\twoheadrightarrow$, $\twoheadrightarrow\!\!\!\!\twoheadrightarrow$, and $\rightarrow\!\!\!\!\times$, which correspond to the ground, activating, and deactivating edges, respectively. The current evidence on whether an edge is active or inactive at any given moment is captured by the activation function $\mathcal{C} : E \to A^2$, while their internal details, as in Definition 5, are given by the functions $\bar{\cdot}$ for ground edges and $\bar{\bar{\cdot}}$ for hyper-edges.

The following definition will be used in the sequel to describe PRG whose values are totally consistent, totally inconsistent, or strictly consistent.

Definition 8. *Let* M *be an* (A, Σ)-PRG *and* $\mathcal{C} : E \to A^2$ *an activation function over the edges of* M*. We say that:*

- M *is* totally consistent *if for every edge* $x \in E$*, it holds that* $\mathcal{C}(x) \in \Delta_C$*.*
- M *is* totally inconsistent *if for every edge* $x \in E$*, it holds that* $\mathcal{C}(x) \in \Delta_P$*.*
- M *is* strictly consistent *if for every edge* $x \in E$*, it holds that* $\mathcal{C}(x) \in \Delta$*.*

Furthermore, since we are interested in consistency-preserving results, the following two theorems regarding Δ_C and Δ_P will be used in the subsequent subsections.

Theorem 1. Δ_C *is closed under the operators* $\neg$, $\Sqcap$, $\Sqcup$ *and* $\otimes$*.*

Proof. Let $(t\!t_1, f\!\!f_1), (t\!t_2, f\!\!f_2) \in \Delta_C$, then

- $\neg(tt_1, ff_1) = (ff_1, tt_1) \in \Delta_C$, that is, we intend to prove that

$$D((ff_1, tt_1), (0,0)) \leqslant D((ff_1, tt_1), (1,1)) \tag{5}$$

By hypothesis $(tt_1, ff_1) \in \Delta_C$, so it follows

$$\begin{aligned} & D((tt_1, ff_1), (0,0)) \leqslant D((tt_1, ff_1), (1,1)) \\ \Leftrightarrow & \sqrt{d(tt_1,0)^2 + d(ff_1,0)^2} \leqslant \sqrt{d(tt_1,1)^2 + d(ff_1,1)^2} \\ \Leftrightarrow & \sqrt{d(ff_1,0)^2 + d(tt_1,0)^2} \leqslant \sqrt{d(ff_1,1)^2 + d(tt_1,1)^2} \\ \Leftrightarrow & D((ff_1, tt_1), (0,0)) \leqslant D((ff_1, tt_1), (1,1)) \end{aligned}$$

Hence, if $(tt_1, ff_1) \in \Delta_C$, then $(ff_1, tt_1) \in \Delta_C$.

- $(tt_1, ff_1) \sqcap\!\!\!\sqcap (tt_2, ff_2) = (tt_1 \sqcap tt_2, ff_1 \sqcup ff_2) \in \Delta_C$, we intend to prove that

$$D((tt_1 \sqcap tt_2, ff_1 \sqcup ff_2), (0,0)) \leqslant D((tt_1 \sqcap tt_2, ff_1 \sqcup ff_2), (1,1)) \tag{6}$$

If $(tt_1 \sqcap tt_2, ff_1 \sqcup ff_2)$ equals either (tt_1, ff_1) or (tt_2, ff_2), then (6) follows immediately from the hypothesis. If $(tt_1 \sqcap tt_2, ff_1 \sqcup ff_2) = (tt_1, ff_2)$. By hypothesis,

$$\begin{aligned} & D((tt_2, ff_2), (0,0)) \leqslant D((tt_2, ff_2), (1,1)) \\ \Leftrightarrow & \sqrt{d(tt_2,0)^2 + d(ff_2,0)^2} \leqslant \sqrt{d(tt',1)^2 + d(ff',1)^2} \\ \Leftrightarrow & d(tt_2,0)^2 + d(ff_2,0)^2 \leqslant d(tt_2,1)^2 + d(ff_2,1)^2 \end{aligned} \tag{7}$$

Since $tt_1 \sqcap tt_2 = tt_1$ and $ff_1 \sqcup ff_2 = ff_2$ it follows $tt_1 \leqslant tt_2$ and $ff_1 \leqslant ff_2$. Thus, $d(tt_1, 0) \leqslant d(tt_2, 0)$ and $d(tt_1, 1) \geqslant d(tt_2, 1)$. Resorting again to (7), it follows

$$\begin{aligned} D((tt_1, ff_2), (0,0)) &= d(tt_1,0)^2 + d(ff_2,0)^2 \\ &\leqslant d(tt_2,0)^2 + d(ff_2,0)^2 \\ &\leqslant d(tt_2,1)^2 + d(ff_2,1)^2 \\ &\leqslant d(tt_1,1)^2 + d(ff_2,1)^2 = D((tt_1, ff_2), (1,1)) \end{aligned}$$

Similarly, it is possible to prove the case $(tt_1 \sqcap tt_2, ff_1 \sqcup ff_2) = (tt_2, ff_1)$.

- $(tt_1, ff_1) \sqcup\!\!\!\sqcup (tt_2, ff_2) \in \Delta_C$. Since $(tt_1, ff_1) \sqcup\!\!\!\sqcup (tt_2, ff_2) = \neg(\neg(tt_1, ff_1) \sqcap\!\!\!\sqcap \neg(tt_2, ff_2))$ it follows from the fact that Δ_C is closed under the operators $\neg$ and $\sqcap\!\!\!\sqcap$.

- $(tt_1, ff_1) \otimes (tt_2, ff_2) = (tt_1 \sqcap tt_2, ff_1 \sqcap ff_2) \in \Delta_C$. If $(tt_1 \sqcap tt_2, ff_1 \sqcap ff_2)$ equals either (tt_1, ff_1) or (tt_2, ff_2), then it follows immediately from the hypothesis. If $(tt_1 \sqcap tt_2, ff_1 \sqcap ff_2) = (tt_1, ff_2)$. By hypothesis,

$$\begin{aligned} & D((tt_1, ff_1), (0,0)) \leqslant D((tt_1, ff_1), (1,1)) \\ \Leftrightarrow & d(tt_1,0)^2 + d(ff_1,0)^2 \leqslant d(tt_1,1)^2 + d(ff_1,1)^2 \end{aligned} \tag{8}$$

Since $tt_1 \sqcap tt_2 = tt_1$ and $ff_1 \sqcap ff_2 = ff_2$ it follows $d(ff_1, 0) \geqslant d(ff_2, 0)$ and $d(ff_1, 1) \leqslant d(ff_2, 1)$. By also using (8), it follows

$$\begin{aligned} D((tt_1, ff_2), (0,0)) &= d(tt_1, 0)^2 + d(ff_2, 0)^2 \\ &\leqslant d(tt_1, 0)^2 + d(ff_1, 0)^2 \\ &\leqslant d(tt_1, 1)^2 + d(ff_1, 1)^2 \\ &\leqslant d(tt_1, 1)^2 + d(ff_2, 1)^2 = D((tt_1, ff_2), (1,1)) \end{aligned}$$

which proves $D((tt_1, ff_2), (0,0)) \leqslant D((tt_1, ff_2), (1,1))$. Similarly, it is possible to prove the case $(tt_1 \sqcap tt_2, ff_1 \sqcap ff_2) = (tt_2, ff_1)$.

□

It is worth noting that Δ_C is not closed under the operator $\oplus$.

For example, let **G** be the underlying lattice. Then, $(0.6, 0.3), (0.3, 0.6) \in \Delta_C$ however, $(0.6, 0.3) \oplus (0.3, 0.6) = (\max\{0.6, 0.3\}, \max\{0.3, 0.6\}) = (0.6, 0.6) \notin \Delta_C$.

The proof of the next Theorem is omitted, as it is similar to that of Theorem 1.

Theorem 2. *Δ_P is closed under the operators $\neg$, $\sqcap\!\!\!\sqcap$, $\sqcup\!\!\!\sqcup$ and $\oplus$.*

Δ_P is not closed under the operator $\otimes$.

For example, let **G** be the underlying lattice. Then, $(0.9, 0.2), (0.2, 0.9) \in \Delta_P$ however, $(0.9, 0.2) \otimes (0.2, 0.9) = (\min\{0.9, 0.2\}, \min\{0.2, 0.9\}) = (0.2, 0.2) \notin \Delta_P$.

The following example illustrates a PRG.

Example 10. *The following diagram represents a* $([0,1], \{a, b\})$-PRG.

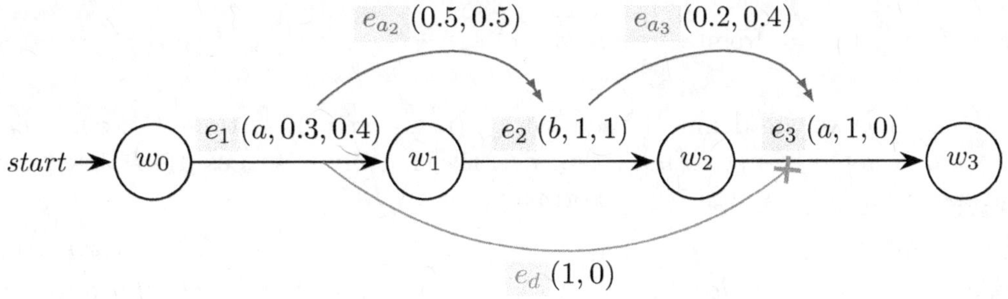

Formally, this PRG *is defined over the set of states* $W = \{w_0, w_1, w_2, w_3\}$, *with initial state* w_0. *The set of ground edges is* $\rightarrow = \{e_1, e_2, e_3\}$, *the set of activating hyper-edges is* $\twoheadrightarrow = \{e_{a_2}, e_{a_3}\}$, *and the set of deactivating hyper-edges is* $\rightarrow\!\!\times = \{e_d\}$.

Each edge $e \in E$ is assigned a pair of truth values by the initial activation function $\mathcal{C}_0 : E \longrightarrow A^2$, as depicted in the diagram.

The functions $\bar{\cdot}$ and $\bar{\bar{\cdot}}$ map ground and hyper-edges, respectively, to their behavior. For example, $\overline{e_1} = (w_0, a, w_1)$ corresponds to the transition associated with ground edge e_1, and $\overline{\overline{e_d}} = (e_1, e_3)$ defines the deactivating hyper-edge e_d that deactivates e_3 upon the execution of e_1.

The following sections address non-conflicting and conflicting hyper-edges, respectively. Before formally defining these notions, we revisit the concepts defined in (1) and (2) with a few changes to accommodate the fact that edges are weighted by pairs of truth values in A^2. This revision is essential to enable a precise definition of what constitutes conflicting versus non-conflicting hyper-edges.

Definition 9. *Let M be an (A, Σ)-PRG where $E = (\rightarrow \cup \twoheadrightarrow \cup \rightarrow\!\!\!\!\times)$ is the set all of edges and $H = (\twoheadrightarrow \cup \rightarrow\!\!\!\!\times)$ is the set of all hyper-edges. For any $e_s, e_t \in E$ we define*

$$\begin{aligned} \mathsf{from}(e_s) &= \{h \in H \mid \exists e_t \in E \;\cdot\; \overline{\overline{h}} = (e_s, e_t)\} \\ \mathsf{to}(e_t) &= \{h \in H \mid \exists e_s \in E \;\cdot\; \overline{\overline{h}} = (e_s, e_t)\} \end{aligned}$$

Intuitively, as before, $\mathsf{from}(e_s)$ returns the set of hyper-edges that directly originate at edge e_s. Similarly, $\mathsf{to}(e_t)$ retrieves all hyper-edges that target edge e_t. We can now inductively define $\mathsf{from}^\star(e_s)$ as the transitive closure of $\mathsf{from}(e_s)$, which returns all hyper-edges that are (directly or indirectly) triggered by traversing e_s.

Definition 10. *Let M be an (A, Σ)-PRG. For any edge $e_s \in E$ we define inductively the set of all hyper-edges triggered by e_s as*

$$\mathsf{from}^\star(e_s) = \bigcup_{n \geqslant 0} \mathsf{from}^{(n)}(e_s)$$

where $\mathsf{from}^{(0)}(e_s) = \mathsf{from}(e_s)$ *and* $\mathsf{from}^{(n)}(e_s) = \bigcup_{h \in \mathsf{from}^{(n-1)}(e_s)} \mathsf{from}(h)$.

As before, we will define the sets $\mathsf{on}(e_s)$ (resp. $\mathsf{off}(e_s)$) as the set of edges $e_t \in E$ that are targeted by activating (resp. deactivating) hyper-edges $h \in H$ triggered by traversing e_s, i.e., $h \in \mathsf{from}^\star(e_s)$. Formally, for a PRG

Definition 11. *Let M be an (A, Σ)-PRG and $e_s \in E$ an edge. The set of edges activated (respectively deactivated) by e_s, written $\mathsf{on}(e_s)$ (respectively, $\mathsf{off}(e_s)$) are defined as follows,*

$$\begin{aligned} \mathsf{on}(e_s) &= \{e_t \in E \mid \exists h \in H \,.\, h \in (\mathsf{from}^\star(e_s) \cap \twoheadrightarrow \cap\, to(e_t))\} \\ \mathsf{off}(e_s) &= \{e_t \in E \mid \exists h \in H \,.\, h \in (\mathsf{from}^\star(e_s) \cap \rightarrow\!\!\!\!\times \cap\, to(e_t))\} \end{aligned} \tag{9}$$

The following example illustrates the definitions introduced above for PRG.

Example 11. *Consider the* $([0,1], \emptyset)$*-PRG depicted below.*[3] *Observe that edge* e_1 *triggers the activating hyper-edge* e_{a_1}*, which in turn triggers the deactivating hyper-edge* e_d*, which itself then triggers the activating hyper-edge* e_{a_2}*.*

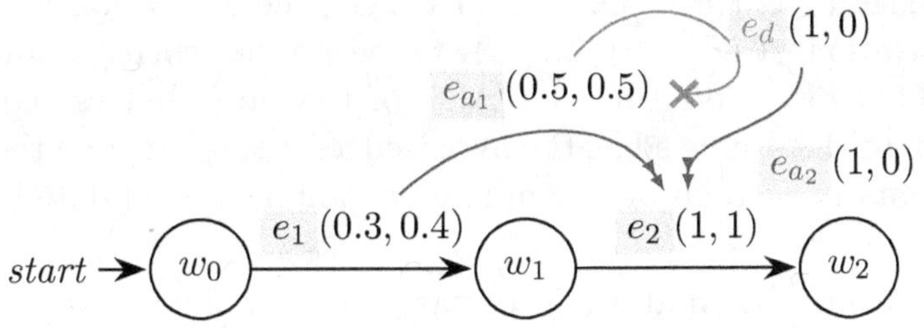

Since $\mathsf{from}(e_1) = \{e_{a_1}\}$, $\mathsf{from}(e_{a_1}) = \{e_d\}$, *and* $\mathsf{from}(e_d) = \{e_{a_2}\}$ *it follows that* $\mathsf{from}^\star(e_1) = \{e_{a_1}, e_{a_2}, e_d\}$*. Next, since* $\mathsf{to}(e_2) = \{e_{a_1}, e_{a_2}\}$ *and* $\mathsf{to}(e_{a_1}) = \{e_d\}$ *one has* $\mathsf{on}(e_1) = \{e_2\}$ *and* $\mathsf{off}(e_1) = \{e_{a_1}\}$*. In other words, as expected, traversing* e_1 *ultimately triggers the hyper-edges* e_{a_1}*,* e_d*, and* e_{a_2}*, which in turn act upon* e_2 *and* e_{a_1} *as targets by activating and deactivating them, respectively.*

Finally, we introduce the notion of a set of hyper-edges adjacent to a pair of edges e_s and e_t.

Definition 12. *Let* M *be an* (A, Σ)*-PRG where* $E = (\rightarrow \cup \twoheadrightarrow \cup \rightarrow\!\!\!\times)$ *is the set all of edges. For any* $e_s, e_t \in E$ *the set* $\Gamma_A(e_s, e_t)$ *(resp.* $\Gamma_D(e_1, e_2)$*) denotes the set of activating (resp. deactivating) hyper-edges that are triggered by traversing edge* e_s *and that target edge* e_t*.*

$$\begin{aligned} \Gamma_A(e_s, e_t) &= \mathsf{from}^\star(e_s) \cap \mathsf{to}(e_t) \cap \twoheadrightarrow \\ \Gamma_D(e_s, e_t) &= \mathsf{from}^\star(e_s) \cap \mathsf{to}(e_t) \cap \rightarrow\!\!\!\times \end{aligned} \tag{10}$$

We write $\Gamma(e_s, e_t) = \Gamma_A(e_s, e_t) \cup \Gamma(e_s, e_t)$ *to denote the set of all hyper-edges that are triggered by traversing* e_s *and that target* e_t*.*

Note that if $h \in \Gamma_A(e_s, e_t)$, then $h \in \mathsf{to}(e_t)$. However, it does not necessarily follow that $h \in \mathsf{from}(e_s)$, since h may be triggered by the traversal of e_s but not directly originate from it (see hyper-edges e_{a_2} and e_d in Example 11). We use the

[3] Note that since the set of action symbols is empty, we omit them from the labels of the ground edges.

projection function $\pi_1 : E \times E \longrightarrow E$, defined by $\pi_1(e_1, e_2) = e_1$, and write $\pi_1(\overline{\overline{h}})$ to denote the edge from which the hyper-edge h originates.

Definition 12 plays a crucial role in the following sections. First, they allow us to formally define when a paraconsistent reactive graph contains no conflicting hyper-edges (see Definition 13). Second, they are fundamental for describing how the system's configuration evolves after traversing an edge.

When an edge e_s is traversed, it will trigger the activation of edges in $\mathsf{on}(e_s)$ and the deactivation of edges in $\mathsf{off}(e_s)$, based on the corresponding hyper-edges in $\Gamma(e_s, e_t)$. Therefore, updating the value of the target edges requires gathering information about the source edge, the hyper-edges it triggers, and the corresponding targets. This data must then be meaningfully combined to update the values of the target edges.

We have now introduced all the necessary tools for the following sections, which focus on three main tasks: formally defining conflicts between hyper-edges, describing how to update the system's configuration when an edge is traversed, and proposing policies for managing conflicts in the presence of contradictory hyper-edges.

4.1 Paraconsistent Reactive Graphs without Conflicts

In this section, we study paraconsistent reactive graphs without conflicting hyper-edges. Intuitively, this means that for any pair of edges, there is at most one hyper-edge connecting them. For instance, the paraconsistent reactive graph in Example 10 contains no conflicting hyper-edges, whereas the one in Example 11 includes a conflict between the activating hyper-edges e_{a_1} and e_{a_2}. Formally,

Definition 13. *Let M be an (A, Σ)-PRG. M is said to have no conflicting hyper-edges if for any $e_s, e_t \in E$, $|\Gamma(e_s, e_t)| \leqslant 1$.*

Intuitively, for any pair of edges e_s, e_t, the set of all hyper-edges triggered by e_s and targeting e_t, denoted by $\Gamma(e_s, e_t)$ as defined in Definition 12, contains at most one element.

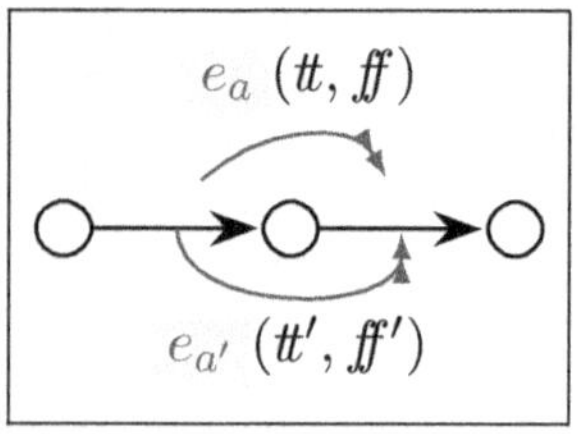

Figure 3

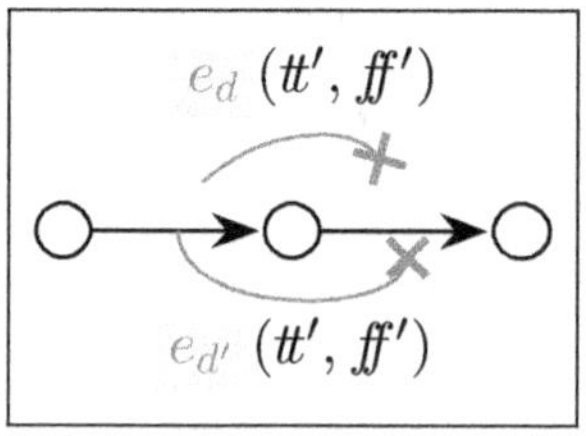

Figure 4

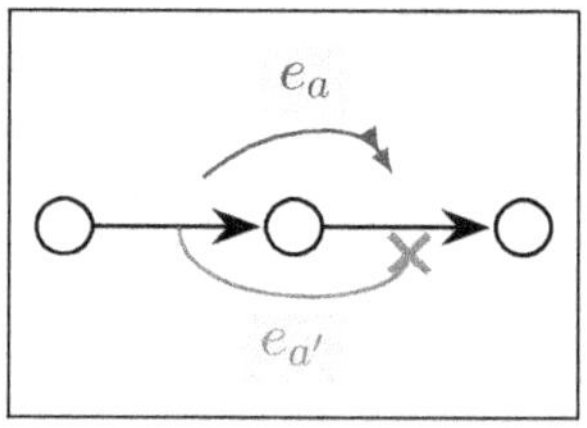

Figure 5

Visually, if $(tt, ff) \neq (tt', ff')$ are truth values in A^2, the scenarios depicted in Figure 3 and Figure 4 are not considered in this section, neither it is the scenario in Figure 5, regardless of the truth values of the edges[4].

It is important to note that in reactive graphs with bivalent hyper-edges (such as those in [11, 30, 42], or the $\mathrm{RG_P}$ introduced in Section 3) conflicting scenarios like those depicted in Figures 3 and 4 do not occur. In such systems, hyper-edges act as deterministic, bivalent switches. Therefore, having two distinct hyper-edges of the same type between the same source and target is redundant.

In contrast, the framework discussed in this paper allows hyper-edges to represent degrees of belief both in favor of and against their effect. As a result, even if two activating (or two deactivating) hyper-edges connect the same source and target, a conflict may arise if their values differ. This raises a natural question: which hyper-edge should take precedence, or how should their effects be combined?

In this section, however, we focus on PRG without conflicts. Our goal is to better understand how individual activating or deactivating hyper-edges influence the system's evolution. The main challenge lies in the fact that hyper-edges are weighted: their effect on a target edge depends not only on their own value, but also on the values of the edge that triggered them and the target edge itself.

More precisely, when an edge e_s is traversed and triggers a hyper-edge e_h targeting an edge e_t, it is necessary to define how to combine the values of e_s, e_h, and e_t to compute the updated value of e_t.

For this purpose, we now introduce two functions, Act and Deact. These functions are applied separately to activating and deactivating hyper-edges, and define how their individual effects impact the system. Note that these are not the only possible choices, but they are particularly simple.

Example 12. *The following functions,* Act *and* Deact*, combine truth values α, β and γ in A^2 into a single value in A^2.*

$$\mathsf{Act}(\alpha,\beta,\gamma) = (\alpha \sqcap\!\!\!\sqcap \beta) \sqcup\!\!\!\sqcup \gamma \qquad \mathsf{Deact}(\alpha,\beta,\gamma) = \neg(\alpha \sqcap\!\!\!\sqcap \beta) \sqcap\!\!\!\sqcap \gamma$$

Let $\alpha \in A^2$ be the value of a traversed edge e_s, let $\beta_a, \beta_d \in A^2$ be the values of the triggered activating and deactivating hyper-edges e_a and e_d, respectively, and let $\gamma \in A^2$ denote the current value of a target edge e_t. There exists two possible scenarios which correspond to the activation of edge e_t, as depicted in Figure 6, or the deactivation of edge e_t, as depicted in Figure 7.

The functions above are defined in order to generalize the expected behavior in the classical bivalent setting, where truth values are either $(1,0)$ *(strictly true) or* $(0,1)$ *(strictly false).*

[4]Note that the black edges depicted on Figures 3, 4 and 5 are not necessarily ground edges.

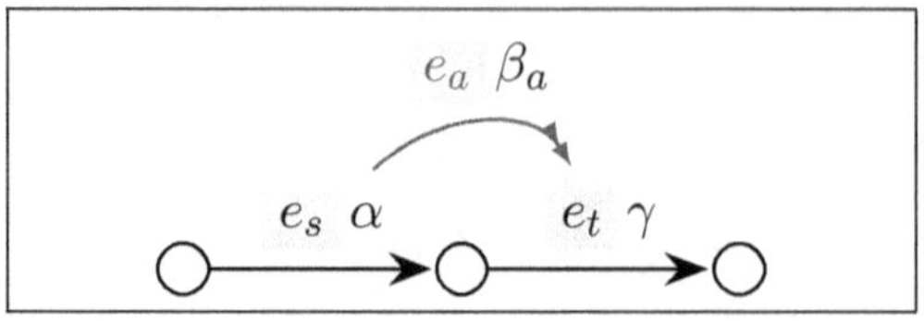

Figure 6

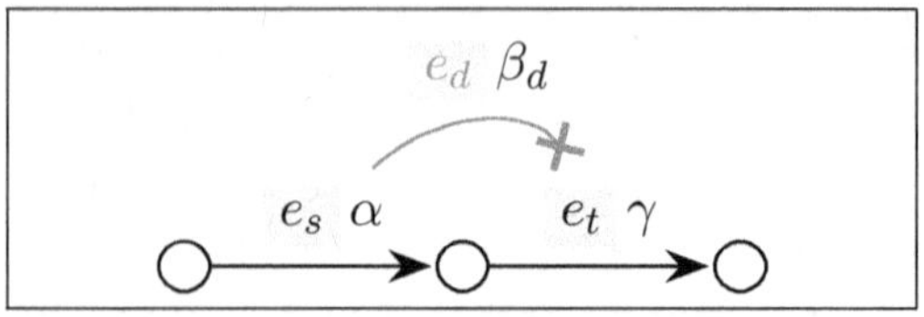

Figure 7

For the activation scenario depicted in Figure 6: when should the target edge e_t be activated? There are two cases to consider:

Case 1: *e_t is already active. In this case, any attempt to activate e_t has no effect. Formally, component $\ldots \sqcup\!\!\!\sqcup \gamma$ in function* Act *ensures that the returned value is at least as great as γ.*

Case 2: *The traversed edge e_s triggers an activating hyper-edge e_a. In this case, the values α and β_a, associated to e_s and e_a, respectively, reinforce the activation of e_t. This is captured by the term $\alpha \sqcap\!\!\!\sqcap \beta_a$ in function* Act.

Similarly, for the deactivation scenario in Figure 7: when should the target edge e_t be deactivated? Again, there are two scenarios to consider:

Case 1: *e_t is already deactivated. In this case, any attempt to deactivate it again has no effect. Formally, component $\ldots \sqcap\!\!\!\sqcap \gamma$ in function* Deact *ensures that the returned value is at best equal to γ.*

Case 2: *The traversed edge e_s triggers a deactivating hyper-edge e_d. In this case, the values α and β_d, associated to e_s and e_d, respectively, negatively reinforce edge e_t. This is captured by $\neg(\alpha \sqcap\!\!\!\sqcap \beta_d)$ in function* Deact.

These cases extend naturally to general values in A^2 via the twist-structure operators employed. As discussed in the sequel, functions Act *and* Deact *define how the values of the traversed edge, the triggered hyper-edge, and the target edge are combined to compute the new value of e_t.*

It is important to highlight that the functions Act and Deact are defined to generalize classic bivalent behavior. That is, if all edges are weighted by the pair $(1, 0)$, that represents absolute truth, or pair $(0, 1)$, that represents absolute falsity, then the output of these functions coincides with the expected result from classical reactive graphs. This important property is formalized in Proposition 1.

It is also important to note that similar challenges have also been addressed in the research on fuzzy reactive graphs [7, 8, 9], where all edges are assigned weights in the interval $[0,1]$. In that setting, the authors use aggregation functions to combine values in $[0,1]$. Formally, an aggregation function is a mapping $\mathsf{A}^{(n)} : [0,1]^n \to [0,1]$ that is monotonic in each variable and satisfies certain boundary conditions [32]. More recent work extended this concept to any bounded poset [21, 36].

However, when edges in a fuzzy reactive graph are restricted to classical truth values, i.e., transitions are either fully possible (1) or not (0)– the output of some aggregation functions do not coincide with the expected bivalent outcomes. For this reason, we refrain from adopting aggregation functions in here to ensure that our framework, when restricted to classical truth values, aligns with the classic theory on reactive graphs.

The following proposition ensures that the functions Act and Deact correspond to the classical activation and deactivation of edges, respectively, when the input truth values are classical, i.e., $(1,0)$ and $(0,1)$. In this setting, if all edges are weighted by the pair $(1,0)$, providing evidence in favour of traversing the edge, or by the pair $(0,1)$, providing the opposite, then the output of these functions coincides with the expected behaviour of classical reactive graphs.

More specifically, as noted in Example 12, an activating edge e_a triggered by a ground edge e_s and targeting an edge e_t, as depicted in Figure 6, results in the activation of e_t either when e_t is already active or when both e_a and e_s are active. Likewise, a deactivating edge e_d triggered by a ground edge e_s and targeting an edge e_t, as shown in Figure 7, results in the deactivation of e_t either when e_t is already inactive or when both e_d and e_s are active. Formally,

Proposition 1. *Let M be an (A,Σ)-PRG with no conflicting hyper-edges such that, for any edge $x \in E$, $\mathcal{C}(x) \in \{(1,0),(0,1)\}$. For any ground edges $e_s, e_t \in \rightarrow$, activating hyper-edge $e_a \in \twoheadrightarrow$, and deactivating hyper-edge $e_d \in \multimap$, it follows that:*

$$\mathsf{Act}(\mathcal{C}(e_s),\mathcal{C}(e_a),\mathcal{C}(e_t)) = \begin{cases} (1,0) & \textit{if } \mathcal{C}(e_t) = (1,0) \textit{ or } \mathcal{C}(e_s) = \mathcal{C}(e_a) = (1,0) \\ (0,1) & \textit{otherwise} \end{cases}$$

$$\mathsf{Deact}(\mathcal{C}(e_s),\mathcal{C}(e_d),\mathcal{C}(e_t)) = \begin{cases} (0,1) & \textit{if } \mathcal{C}(e_t) = (0,1) \textit{ or } \mathcal{C}(e_s) = \mathcal{C}(e_a) = (1,0) \\ (1,0) & \textit{otherwise} \end{cases}$$

Proof. Let M be an (A,Σ)-PRG with no conflicting hyper-edges. Consider two edges $e_s, e_t \in E$, and let $e_a \in \twoheadrightarrow$ and $e_d \in \multimap$ be an activating and a deactivating hyper-edge, respectively. Let α denote the value of the traversed edge e_s, β_a and β_d the values of the triggered hyper-edges e_a and e_d, and γ the current value of the target edge e_t. We consider the two scenarios illustrated in Figures 6 and 7.

If values α, β_a, β_d, and γ are restricted to classical truth values, i.e., $(1,0)$ (true) and $(0,1)$ (false), then the behavior of functions Act and Deact coincides with the classical theory.

We first consider the outputs of $\mathsf{Act}(\alpha, \beta_a, \gamma)$, which give the updated truth value of e_t after traversing e_s in the activation scenario of Figure 6.

If $\alpha = (1,0)$, then

$\mathsf{Act}(\alpha, \beta_a, \gamma)$	$\gamma = (0,1)$	$\gamma = (1,0)$
$\beta_a = (0,1)$	$(0,1)$	$(1,0)$
$\beta_a = (1,0)$	$(1,0)$	$(1,0)$

If $\alpha = (0,1)$, then

$\mathsf{Act}(\alpha, \beta_a, \gamma)$	$\gamma = (0,1)$	$\gamma = (1,0)$
$\beta_a = (0,1)$	$(0,1)$	$(1,0)$
$\beta_a = (1,0)$	$(0,1)$	$(1,0)$

As expected, the edge e_t is activated if and only if either it was already active, i.e., $\gamma = (1,0)$, or if both the traversed edge and the activating hyper-edge are active, i.e., $\alpha = \beta_a = (1,0)$.

Next, we consider $\mathsf{Deact}(\alpha, \beta_d, \gamma)$, which computes the updated value of e_t in the deactivation scenario depicted in Figure 7.

If $\alpha = (1,0)$, then

$\mathsf{Deact}(\alpha, \beta_d, \gamma)$	$\gamma = (0,1)$	$\gamma = (1,0)$
$\beta_d = (0,1)$	$(0,1)$	$(1,0)$
$\beta_d = (1,0)$	$(0,1)$	$(0,1)$

If $\alpha = (0,1)$, then

$\mathsf{Deact}(\alpha, \beta_d, \gamma)$	$\gamma = (0,1)$	$\gamma = (1,0)$
$\beta_d = (0,1)$	$(0,1)$	$(1,0)$
$\beta_d = (1,0)$	$(0,1)$	$(1,0)$

Again, as expected, the edge e_t is deactivated if and only if either it was already inactive, i.e., $\gamma = (0,1)$, or if both the traversed edge and the deactivating hyper-edge are active, i.e., $\alpha = \beta_d = (1,0)$. □

The following proposition studies monotonicity properties of functions Act and Deact introduced in Example 12.

Proposition 2. *Let $\alpha, \alpha', \beta, \beta', \gamma$ and γ' be in A^2. Then the following conditions hold with respect to the truth-order $\leqslant_t$.*

1. *If $\alpha \leqslant_t \alpha'$, $\beta \leqslant_t \beta'$ and $\gamma \leqslant_t \gamma'$, then $\mathsf{Act}(\alpha, \beta, \gamma) \leqslant_t \mathsf{Act}(\alpha', \beta', \gamma')$.*
2. *If $\alpha \leqslant_t \alpha'$, $\beta \leqslant_t \beta'$ and $\gamma \geqslant_t \gamma'$, then $\mathsf{Deact}(\alpha, \beta, \gamma) \geqslant_t \mathsf{Deact}(\alpha', \beta', \gamma')$.*

And the following conditions hold with respect to the knowledge-order $\leqslant_k$.

3. *If $\alpha \leqslant_k \alpha'$, $\beta \leqslant_k \beta'$ and $\gamma \leqslant_k \gamma'$, then $\mathsf{Act}(\alpha, \beta, \gamma) \leqslant_k \mathsf{Act}(\alpha', \beta', \gamma')$.*
4. *If $\alpha \leqslant_k \alpha'$, $\beta \leqslant_k \beta'$ and $\gamma \leqslant_k \gamma'$, then $\mathsf{Deact}(\alpha, \beta, \gamma) \leqslant_k \mathsf{Deact}(\alpha', \beta', \gamma')$.*

Proof. The proof relies on well-known properties of twist-structures and the associated orders $\leqslant_t$ and $\leqslant_k$. In particular, we use the fact that De Morgan laws hold, e.g., $\neg(\alpha \Cap \beta) = \neg\alpha \Cup \neg\beta$ (see [19, Lemma 2]). Also, since twist-structures form (square) bilattices (see [1, Definition 2.1]), it follows that:

$$\text{if } \alpha \leqslant_t \alpha', \text{ then } \neg\alpha \geqslant_t \neg\alpha' \tag{11}$$

$$\text{if } \alpha \leqslant_k \alpha', \text{ then } \neg\alpha \leqslant_k \neg\alpha' \tag{12}$$

The conditions 1 and 3 follow immediately from the monotonicity of $\sqcap$ and $\sqcup$ in the underlying lattice $(A, \leqslant)$, which lifts to the monotonicity of $\Cap$ and $\Cup$ with respect to $\leqslant_t$ and $\leqslant_k$.

To prove condition 2 we start with the hypothesis $\alpha \leqslant_t \alpha'$ and $\beta \leqslant_t \beta'$. It follows from (11) and the monotonicity of $\Cup$ that $\neg\alpha \Cup \neg\beta \geqslant_t \neg\alpha' \Cup \neg\beta'$. By De Morgan, this is equivalent to $\neg(\alpha \Cap \beta) \geqslant_t \neg(\alpha' \Cap \beta')$. Since $\gamma \geqslant_t \gamma'$ and $\Cap$ is monotonic, we conclude $\neg(\alpha \Cap \beta) \Cap \gamma \geqslant_t \neg(\alpha' \Cap \beta') \Cap \gamma'$, that is, $\mathsf{Deact}(\alpha, \beta, \gamma) \geqslant_t \mathsf{Deact}(\alpha', \beta', \gamma')$.

Similarly, to prove 4 we start with the hypothesis $\alpha \leqslant_k \alpha'$ and $\beta \leqslant_k \beta'$. It follows from (12), the monotonicity of $\Cup$ and De Morgan laws that $\neg\alpha \Cup \neg\beta \leqslant_k \neg\alpha' \Cup \neg\beta'$. Finally, since $\gamma \leqslant_k \gamma'$ and by the monotonicity of $\Cap$, we prove that $\mathsf{Deact}(\alpha, \beta, \gamma) \leqslant_k \mathsf{Deact}(\alpha', \beta', \gamma')$. □

Informally, condition 1 in the previous proposition states that the activation function, Act, is monotonic in the truth order: the stronger the evidence for the traversed edge, the activating hyper-edge, and the target edge, the stronger the activation effect. Condition 2 describes the expected behavior of the deactivation function, Deact, the higher the belief in the traversed and deactivating edges, the more the function impacts the target towards deactivation (i.e., a lower output), and the more active the target edge is initially, the less impact the deactivation will have. Conditions (3) and (4) confirm that both Act and Deact are monotonic with respect to the knowledge order ($\leqslant_k$), i.e., more information in the inputs leads to more information in the output.

As in Section 3, we now aim at defining a semantics for "unfolding" PRG with no conflicting hyper-edges into a paraconsistent graph. For this, we first clarify how the activation function of a PRG changes after traversing an edge, using the functions Act and Deact introduced in Definition 12.

Definition 14. *Given an (A, Σ)-paraconsistent reactive graph with non conflicting hyper-edges let $\mathcal{C} : E \to A^2$ be a activation function and $e, x \in E$, then*

$$\mathcal{C}^e(x) = \begin{cases} \mathcal{C}(x) & \text{if } x \notin (\mathsf{on}(e) \cup \mathsf{off}(e)) \\ \mathsf{Act}(\mathcal{C}(\pi_1(\overline{\overline{h}})), \mathcal{C}(h), \mathcal{C}(x)) & \text{if } \exists h \in H \cdot h \in \Gamma_A(e, x) \\ \mathsf{Deact}(\mathcal{C}(\pi_1(\overline{\overline{h}})), \mathcal{C}(h), \mathcal{C}(x)) & \text{if } \exists h \in H \cdot h \in \Gamma_D(e, x) \end{cases}$$

is the updated value of the edge x, i.e. $\mathcal{C}(x)$, after transvesing e.

In other words, if edge x is not affected by any hyper-edge triggered by e, its value remains unchanged. Otherwise, since we assume no conflicting hyper-edges, there exists a unique hyper-edge h triggered by e[5] that targets x. If h is activating (resp. deactivating), then the updated value of x is computed using Act (resp. Deact), based on the current values of $\pi_1(\overline{\overline{h}})$, h, and x, given by the current activation function $\mathcal{C}$.

We are now equipped to define the semantics for PRG without conflicting hyper-edges. As before, we begin from an initial configuration $\langle w_0, \mathcal{C}_0\rangle$. Then, for any ground edge $e \in \twoheadrightarrow$ going out of state w_0, we compute the updated activation function $\mathcal{C}^e(x)$ for all edges x according to Definition 14.

Definition 15. *The semantics of an (A, Σ)-paraconsistent graph P is given by evolving a configuration $\langle w, \mathcal{C}\rangle$, with $w \in W$ and $\mathcal{C} : E \to A^2$, starting from $\langle w_0, \mathcal{C}_0\rangle$, as defined by the following rule.*

$$\frac{\exists e \in \twoheadrightarrow \cdot \ \overline{e} = (w, a, w')}{\langle w, \mathcal{C}\rangle \xrightarrow{a,\, \mathcal{C}(e)}_P \langle w', \mathcal{C}^e\rangle}$$

In practice, we are only interested in transversing ground edges $e \in \twoheadrightarrow$ such that $\mathcal{C}(e) \neq (0, 1)$, which corresponds to edges that are not inactive.

Let us apply the semantics defined above to the PRG in Example 10.

Example 13. *Consider the* $([0, 1], \{a, b\})$-PRG *without conflicting hyper-edges from Example 10, and the following* $([0, 1], \{a, b\})$*-paraconsistent graph P:*

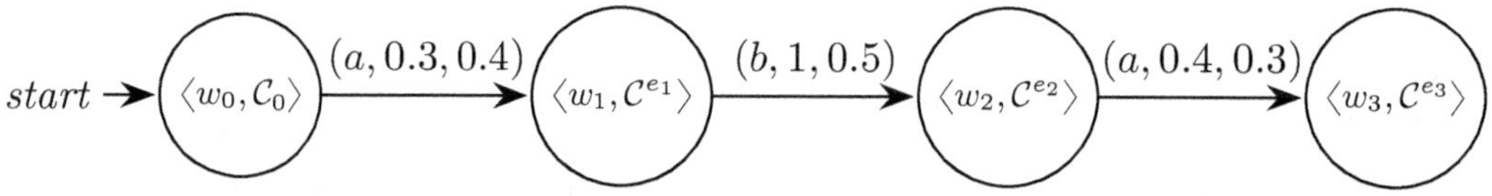

We will show that unfolding the PRG *from Example 10, according to the semantics defined in Definition 15, yields the standard paraconsistent graph P.*

The initial configuration is $\langle w_0, \mathcal{C}_0\rangle$, where the current state is w_0, colored purple, and $\mathcal{C}_0$ assigns to each edge in E the values as depicted below.

[5] Recall that $\pi_1(\overline{\overline{h}})$ denotes the edge from which the hyper-edge h originates. While h is triggered by e_s, it does not necessarily originate directly from e_s.

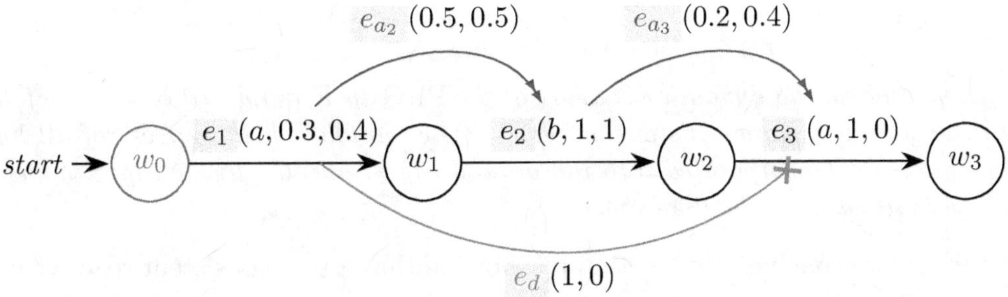

There exists a ground edge e_1 *such that* $\overline{e_1} = (w_0, a, w_1)$. *We now compute for any* $x \in E$ *the updated activation function* $\mathcal{C}^{e_1}(x)$ *after traversing* e_1. *Since* $\mathsf{from}^\star(e_1) = \{e_{a_2}, e_d\}$ *and* $\mathsf{on}(e_1) \cup \mathsf{off}(e_1) = \{e_2, e_3\}$, *the updated activation function is:*

$$\mathcal{C}^{e_1}(x) = \begin{cases} \mathcal{C}(x) & \text{if } x \in E \backslash \{e_2, e_3\} \\ (1, 0.5) & \text{if } x = e_2 \\ (0.4, 0.3) & \text{if } x = e_3 \end{cases}$$

This corresponds to the transition $\langle w_0, \mathcal{C}_0 \rangle \xrightarrow{(a,0.3,0.4)}_P \langle w_1, \mathcal{C}^{e_1} \rangle$ *in the paraconsistent graph* P. *After transversing edge* e_1, *the updated* PRG *in 10 evolves to*

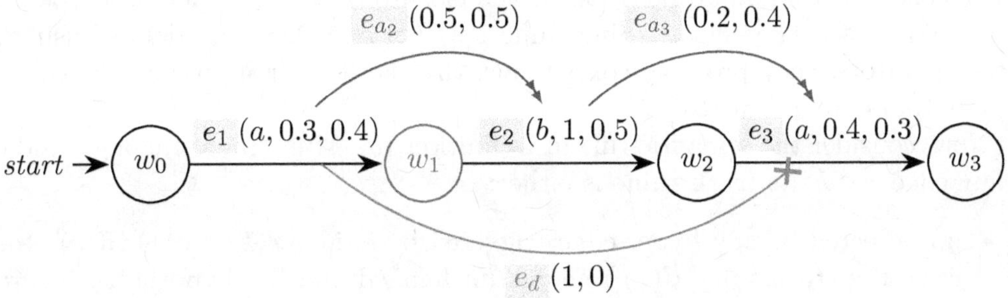

The system is now at state w_1, *colored purple, and there exists* e_2 *such that* $\overline{e_2} = (w_1, b, w_2)$. *Since* $\mathsf{from}^\star(e_2) = \{e_{a_3}\}$ *and* $\mathsf{on}(e_2) \cup \mathsf{off}(e_2) = \{e_3\}$, *the updated activation function is:*

$$\mathcal{C}^{e_2}(x) = \begin{cases} \mathcal{C}^{e_1}(x) & \text{if } x \in E \backslash \{e_3\} \\ (0.4, 0.3) & \text{if } x = e_3 \end{cases}$$

This results in the transition $\langle w_1, \mathcal{C}^{e_1} \rangle \xrightarrow{(1,0.5)}_P \langle w_2, \mathcal{C}^{e_2} \rangle$. *Finally, traversing* e_3 *leads to no further updates, since it triggers no hyper-edges, i.e.,* $\mathcal{C}^{e_3}(x) = \mathcal{C}^{e_2}(x)$

for any $x \in E$*. Thus, this corresponds to the transition* $\langle w_2, \mathcal{C}^{e_2}\rangle \xrightarrow{(0.4,0.3)}_P \langle w_3, \mathcal{C}^{e_3}\rangle$ *in the paraconsistent graph* P*.*

In summary, the dynamic behavior of the PRG *in Example 10 can be "unfolded" into the standard paraconsistent graph* P*. After unfolding, the effects of all hyper-edges have been incorporated into the ground edges, and the hyper-edges themselves may be omitted from the diagram.*

Finally, we conclude this section by showing that a paraconsistent reactive graph without conflicting hyper-edges maintains consistency throughout its evolution, as long as all initial ground and hyper-edges are not weighted by inconsistent truth values in A^2. More specifically, if M is a PRG without conflicting hyper-edges and it is totally consistent, then the system evolves without ever generating inconsistency.

Proposition 3. *Let* M *be a* PRG *without conflicting hyper-edges. If* M *is totally consistent, that is,* $\mathcal{C}_0(x) \in \Delta_C$ *for every edge* $x \in E$*, then the system evolves from the initial configuration by traversing ground edges* $e_1, \ldots, e_n$ *to subsequent configurations, as given by Definition 15, written as*

$$\langle w_0, C_0\rangle \xrightarrow{a\in\Sigma,\, \mathcal{C}(e_1)}_P \langle w_1, C_1\rangle \xrightarrow{a\in\Sigma,\, \mathcal{C}(e_2)}_P \cdots \xrightarrow{a\in\Sigma,\, \mathcal{C}(e_n)}_P \langle w_n, C_n\rangle.$$

and for every edge $x \in E$*, it follows that* $C_1(x), \ldots, C_n(x) \in \Delta_C$*.*

Proof. The proof follows directly from the closure of set Δ_C under the operations $\neg$, $\Cap$, and $\Cup$ (see Theorem 1). Since functions Act and Deact are defined using only these operators, they preserve consistency, that is, if their inputs are in Δ_C, then their outputs are also in Δ_C.

Now, consider the semantic rule in Definition 15. After traversing a ground edge e, any edge $x \in E$ its truth value is either:

- not affected by any hyper-edge triggered by e, i.e., $x \notin \mathsf{on}(e) \cup \mathsf{off}(e)$. In this case, the truth value $\mathcal{C}(x)$ remains unchanged and, by hypothesis, is already in Δ_C, or
- updated using Act or Deact functions, since the inputs are themselves in Δ_C, the updated value of $\mathcal{C}(x)$ is also in Δ_C.

□

4.2 Paraconsistent Reactive Graphs with Conflicts

In this section, we explore policies for managing conflicting hyper-edges in paraconsistent reactive graphs. Recall from Definition 13 that a paraconsistent reactive graph M, as defined in Definition 7, has conflicting hyper-edges if there exist edges $e_s, e_t \in E$ such that $|\Gamma(e_s, e_t)| > 1$, that is, more than one hyper-edge is triggered by e_s and targets e_t.

In the previous section (Section 4.1), we restricted our attention to paraconsistent reactive graphs whose hyper-edges could be internally inconsistent but never conflicted with one another. Moreover, we showed in Proposition 3 that as long as the system starts with a consistent activation function, it evolves without ever becoming inconsistent.

We now lift this restriction to the general case in which multiple, potentially contradictory hyper-edges target the same edge. This allows us to address scenarios such as those depicted in Figures 3, 4 and 5. This challenge raises fundamental questions, such as: How should we combine the information provided by multiple hyper-edges? Should certain hyper-edges take precedence over others? If so, which policies should guide such choices of precedence?

In bivalent switch graphs resolving such conflicts is straightforward: one simply chooses to prioritize either activation or deactivation [42]. In this section, however, rather than relying solely on binary resolution policies, we consider more flexible approaches that allow multiple hyper-edges to contribute their beliefs. For this purpose, we introduce different functions that combine the information provided by all relevant hyper-edges into a single representative value. Each one is guided by a specific conflict resolution policy that motivates and justifies the chosen strategy.

Let us start by introducing functions $\mathsf{gather}^{\mathsf{A}}_{\odot}$ and $\mathsf{gather}^{\mathsf{D}}_{\odot}$ which will be used to combine the information provided by multiple conflicting activating and deactivating hyper-edges, respectively.

Definition 16. *Let M be any (A, Σ)-PRG where $E = (\twoheadrightarrow \cup \twoheadrightarrow\!\!\!\!\rightarrow \cup \rightarrow\!\!\!\!\times)$ is the set of all edges. For $\odot \in \{\sqcap\!\!\!\sqcap, \sqcup\!\!\!\sqcup\}$ the function*

$$\begin{aligned}\mathsf{gather}^{A}_{\odot} : \ & E \times E \longrightarrow A^2 \\ & e_s, e_t \longmapsto \bigodot_{h \in \Gamma_A(e_s, e_t)} \mathsf{Act}(\mathcal{C}(\pi_1(\overline{\overline{h}})), \mathcal{C}(h), \mathcal{C}(e_t))\end{aligned}$$

computes the updated value of e_t by combining the information of all activating hyper-edges triggered by e_s and that target e_t and the function

$$\begin{aligned}\mathsf{gather}^{D}_{\odot} : \ & E \times E \longrightarrow A^2 \\ & e_s, e_t \longmapsto \bigodot_{h \in \Gamma_A(e_s, e_t)} \mathsf{Deact}(\mathcal{C}(\pi_1(\overline{\overline{h}})), \mathcal{C}(h), \mathcal{C}(e_t))\end{aligned}$$

computes the updated value of e_t by combining the information of all deactivating hyper-edges triggered by e_s and that target e_t.

Intuitively, $\mathsf{gather}^A_{\Cup}$ and $\mathsf{gather}^D_{\Cup}$ correspond to a "best-case" scenario: the value of e_t is updated to the highest possible result among all contributions. Conversely, $\mathsf{gather}^A_{\Cap}$ and $\mathsf{gather}^D_{\Cap}$ represent a "worst-case" scenario, where the lowest result is chosen.

For example, if there are conflicting activating hyper-edges targeting e_t, then $\mathcal{C}(e_t)$ is updated to $(1,0)$ by $\mathsf{gather}^A_{\Cup}$ if at least one hyper-edge $h \in \Gamma_A(e_s, e_t)$ completely supports activation, i.e., $\mathsf{Act}(\mathcal{C}(\pi_1(\bar{\bar{h}})), \mathcal{C}(h), \mathcal{C}(e_t)) = (1,0)$. On the other hand, $\mathsf{gather}^A_{\Cap}$ requires that all hyper-edges in $\Gamma_A(e_s, e_t)$ support complete activation of e_t.

Similarly, if there are conflicting deactivating hyper-edges targeting e_t, then e_t is completely deactivated by $\mathsf{gather}^D_{\Cup}$, i.e. $\mathcal{C}(e_t)$ is updated to $(0,1)$, only if every $h \in \Gamma_D(e_s, e_t)$ supports complete deactivation of e_t, i.e., $\mathsf{Deact}(\mathcal{C}(\pi_1(\bar{\bar{h}})), \mathcal{C}(h), \mathcal{C}(e_t)) = (0,1)$. By contrast, $\mathsf{gather}^D_{\Cap}$ requires only one such hyper-edge to support complete deactivation of e_t.

Of course the choice of operator $\odot$ depends on the problem at hand. Informally, choosing $\Cap$ reflects:

- a conservative stance for activation, requiring consensus among hyper-edges;
- a pessimistic stance for deactivation, where limited evidence suffices to deactivate.

In contrast, using $\Cup$ corresponds to:

- a lenient stance for activation, where limited evidence is sufficient;
- an optimistic stance for deactivation, where full consensus is needed to justify deactivation.

Thus, if we face a conflict between hyper-edges of the same type, as in Figure 3 and Figure 4, we can use $\mathsf{gather}^A_{\odot}$ and $\mathsf{gather}^D_{\odot}$, respectively, to update the value of the target edge, depending on whether we value any available evidence or require consensus among all hyper-edges.

The following two examples illustrate the different conflict resolution policies defined above.

Example 14. *Consider the following* $([0,1], \emptyset)$-PRG *with conflicted hyper-edges.*

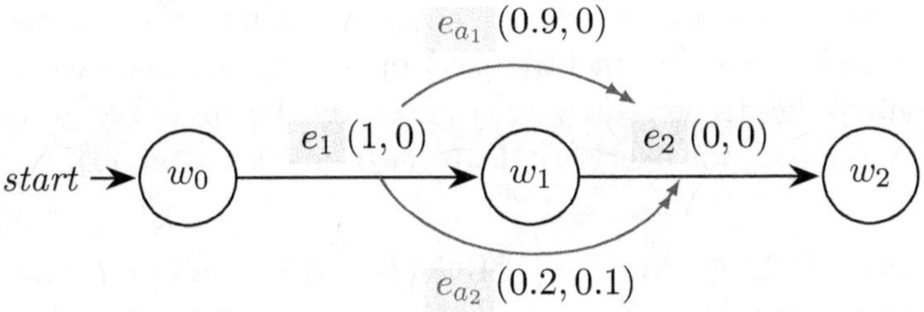

Consequently,

$$\begin{aligned}\mathsf{Act}(\mathcal{C}(e_1), \mathcal{C}(e_{a_1}), \mathcal{C}(e_2)) &= ((1,0) \sqcap (0.9,0)) \sqcup (0,0) = (0.9,0)\\ \mathsf{Act}(\mathcal{C}(e_1), \mathcal{C}(e_{a_2}), \mathcal{C}(e_2)) &= ((1,0) \sqcap (0.2,0.1)) \sqcup (0,0) = (0.2,0)\end{aligned}$$

Since $(0.2, 0.1) \leqslant_t (0.9, 0)$, *i.e.* $\mathcal{C}(e_{a_2}) \leqslant_t \mathcal{C}(e_{a_1})$, *it follows from Proposition 2 it follows that* $\mathsf{Act}(\mathcal{C}(e_1), \mathcal{C}(e_{a_2}), \mathcal{C}(e_2)) \leqslant_t \mathsf{Act}(\mathcal{C}(e_1), \mathcal{C}(e_{a_1}), \mathcal{C}(e_2))$.

Additionally, we have $\mathsf{gather}^{A}_{\sqcap}(e_1, e_2) = (0.2, 0)$ *and* $\mathsf{gather}^{A}_{\sqcup}(e_1, e_2) = (0.9, 0)$ *which shows that using* $\sqcap$ *to combine the information from the activating hyper-edges results in a more conservative (weaker) activation of* e_2, *whereas using* $\sqcup$ *yields a more lenient (stronger) activation.*

Example 15. *Consider the following* $([0,1], \emptyset) - \mathrm{PRG}$ *with conflicted hyper-edges.*

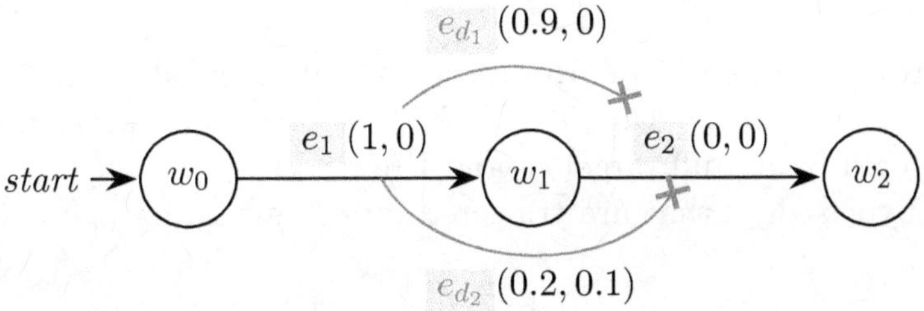

Since $\mathcal{C}(e_{d_2}) \leqslant_t \mathcal{C}(e_{d_1})$, *it follows from Proposition 2 that*

$$(0, 0.2) = \mathsf{Deact}(\mathcal{C}(e_1), \mathcal{C}(e_{d_2}), \mathcal{C}(e_2)) \geqslant_t \mathsf{Deact}(\mathcal{C}(e_1), \mathcal{C}(e_{d_1}), \mathcal{C}(e_2)) = (0, 0.9)$$

Intuitively, hyper-edge e_{d_1} *has a higher truth value than* e_{d_2} *so it produces a stronger deactivation effect on edge* e_2.

Additionally, $\mathsf{gather}^{D}_{\sqcap}(e_1, e_2) = (0, 0.9)$ *and* $\mathsf{gather}^{D}_{\sqcup}(e_1, e_2) = (0, 0.2)$, *which illustrates that using the operator* $\sqcap$ *results in a more pessimistic stance, pushing the value closer to strictly false* $(0, 1)$, *while* $\sqcup$ *leads to a more optimistic outcome, with a weaker deactivation.*

The following proposition provides a simplification in the calculation of $\mathsf{gather}^A_\odot$ and $\mathsf{gather}^D_\odot$ if all hyper-edges in $\Gamma_A(e_s, e_t)$ or $\Gamma_D(e_s, e_t)$ originate directly from e_s. That is, no hyper-edge triggered by e_s triggers another hyper-edge that also targets e_t. Formally, $\pi_1(\overline{\overline{h}}) = e_s$ for every $h \in \Gamma_A(e_s, e_t)$ or $h \in \Gamma_D(e_s, e_t)$.

Proposition 4. *Let M be an (A, Σ)-PRG. For $e_s, e_t \in E$, such that $\Gamma_A(e_s, e_t) \neq \emptyset$, if $\pi_1(\overline{\overline{h}}) = e_s$ for every $h \in \Gamma_A(e_s, e_t)$, then*

$$\bigsqcup_{h \in \Gamma_A(e_s, e_t)} \mathsf{Act}(\mathcal{C}(e_s), \mathcal{C}(h), \mathcal{C}(e_t)) = \mathsf{Act}\left(\mathcal{C}(e_s), \bigsqcup_{h \in \Gamma_A(e_s, e_t)} \mathcal{C}(h), \mathcal{C}(e_t)\right) \tag{13}$$

$$\bigsqcap_{h \in \Gamma_A(e_s, e_t)} \mathsf{Act}(\mathcal{C}(e_s), \mathcal{C}(h), \mathcal{C}(e_t)) = \mathsf{Act}\left(\mathcal{C}(e_s), \bigsqcap_{h \in \Gamma_A(e_s, e_t)} \mathcal{C}(h), \mathcal{C}(e_t)\right) \tag{14}$$

For $e_s, e_t \in E$, such that $\Gamma_D(e_s, e_t) \neq \emptyset$, if $\pi_1(\overline{\overline{h}}) = e_s$ for every $h \in \Gamma_D(e_s, e_t)$, then

$$\bigsqcap_{h \in \Gamma_D(e_s, e_t)} \mathsf{Deact}(\mathcal{C}(e_s), \mathcal{C}(h), \mathcal{C}(e_t)) = \mathsf{Deact}\left(\mathcal{C}(e_s), \bigsqcup_{h \in \Gamma_D(e_s, e_t)} \mathcal{C}(h), \mathcal{C}(e_t)\right) \tag{15}$$

$$\bigsqcup_{h \in \Gamma_D(e_s, e_t)} \mathsf{Deact}(\mathcal{C}(e_s), \mathcal{C}(h), \mathcal{C}(e_t)) = \mathsf{Deact}\left(\mathcal{C}(e_s), \bigsqcap_{h \in \Gamma_D(e_s, e_t)} \mathcal{C}(h), \mathcal{C}(e_t)\right) \tag{16}$$

Proof. We prove (13) and (15). Results (14) and (16) follow similarly. For some $m, n \in \mathbb{N}$, let $\Gamma_A(e_s, e_t) = \{h^A_1, h^A_2, \ldots, h^A_m\}$ be the set of activating hyper-edges that are triggered by e_s and target e_t and $\Gamma_D(e_s, e_t) = \{h^D_1, h^D_2, \ldots, h^D_n\}$ the set of deactivating hyper-edges that are triggered by e_s and target e_t. The proof of (13) is as follows

$$\begin{aligned}
&\bigsqcup_{h \in \Gamma_A(e_s, e_t)} \mathsf{Act}(\mathcal{C}(e_s), \mathcal{C}(h), \mathcal{C}(e_t)) \\
&= \mathsf{Act}(\mathcal{C}(e_s), \mathcal{C}(h^A_1), \mathcal{C}(e_t)) \sqcup \ldots \sqcup \mathsf{Act}(\mathcal{C}(e_s), \mathcal{C}(h^A_m), \mathcal{C}(e_t)) && \{\text{defn. of } \mathsf{Act}\} \\
&= ((\mathcal{C}(e_s) \sqcap \mathcal{C}(h^A_1)) \sqcup \mathcal{C}(e_t)) \sqcup \ldots \sqcup ((\mathcal{C}(e_s) \sqcap \mathcal{C}(h^A_m)) \sqcup \mathcal{C}(e_t)) && \{\sqcup \text{ is involutive}\} \\
&= (\mathcal{C}(e_s) \sqcap \mathcal{C}(h^A_1)) \sqcup \ldots \sqcup (\mathcal{C}(e_s) \sqcap \mathcal{C}(h^A_m)) \sqcup \mathcal{C}(e_t) && \{\text{dist. of } \sqcap \text{ over } \sqcup\} \\
&= \left(\mathcal{C}(e_s) \sqcap \bigsqcup_{h \in \Gamma_A(e_s, e_t)} \mathcal{C}(h)\right) \sqcup \mathcal{C}(e_t) && \{\text{defn. of } \mathsf{Act}\} \\
&= \mathsf{Act}\left(\mathcal{C}(e_s), \bigsqcup_{h \in \Gamma_A(e_s, e_t)} \mathcal{C}(h), \mathcal{C}(e_t)\right)
\end{aligned}$$

The proof of (15) is follows as

$$
\begin{aligned}
&\bigsqcap_{h \in \Gamma_D(e_s,e_t)} \mathsf{Deact}(\mathcal{C}(e_s), \mathcal{C}(h), \mathcal{C}(e_t)) \\
&= \mathsf{Deact}(\mathcal{C}(e_s), \mathcal{C}(h_1^D), \mathcal{C}(e_t)) \Sqcap \ldots \Sqcap \mathsf{Deact}(\mathcal{C}(e_s), \mathcal{C}(h_n^D), \mathcal{C}(e_t)) && \{\text{defn. of } \mathsf{Deact}\} \\
&= (\neg(\mathcal{C}(e_s) \Sqcap \mathcal{C}(h_1^D)) \Sqcap \mathcal{C}(e_t)) \Sqcap \ldots \Sqcap (\neg(\mathcal{C}(e_s) \Sqcap \mathcal{C}(h_n^D)) \Sqcap \mathcal{C}(e_t)) && \{\Sqcap \text{ is involutive}\} \\
&= \neg(\mathcal{C}(e_s) \Sqcap \mathcal{C}(h_1^D)) \Sqcap \ldots \Sqcap \neg(\mathcal{C}(e_s) \Sqcap \mathcal{C}(h_n^D)) \Sqcap \mathcal{C}(e_t) && \{\text{De Morgan law}\} \\
&= (\neg\mathcal{C}(e_s) \Sqcup \neg\mathcal{C}(h_1^D)) \Sqcap \ldots \Sqcap (\neg\mathcal{C}(e_s) \Sqcup \neg\mathcal{C}(h_n^D)) \Sqcap \mathcal{C}(e_t) && \{\text{dist. of } \Sqcup \text{ over } \Sqcap\} \\
&= \left(\neg\mathcal{C}(e_s) \Sqcup \bigsqcap_{h \in \Gamma_D(e_s,e_t)} \neg\mathcal{C}(h)\right) \Sqcap \mathcal{C}(e_t) && \{\text{De Morgan law}\} \\
&= \neg\left(\mathcal{C}(e_s) \Sqcap \bigsqcup_{h \in \Gamma_D(e_s,e_t)} \mathcal{C}(h)\right) \Sqcap \mathcal{C}(e_t) && \{\text{defn of } \mathsf{Deact}\} \\
&= \mathsf{Deact}\left(\mathcal{C}(e_s), \bigsqcup_{h \in \Gamma_D(e_s,e_t)} \mathcal{C}(h), \mathcal{C}(e_t)\right)
\end{aligned}
$$

□

The following corollary is immediate from the fact that Δ_C is closed under operations $\neg$, $\Sqcap$ and $\Sqcup$ as shown in Theorem 1. As before, it ensures that if all edges in the system are initially not inconsistent, then their combination does not produce inconsistent values.

Corollary 1. *Let M be any (A, Σ)-PRG. If M is totally consistent, then, for $e_s, e_t \in E$ and $\odot \in \{\Sqcap, \Sqcup\}$ it follows that* $\mathsf{gather}^A_\odot$, $\mathsf{gather}^D_\odot(e_1, e_2) \in \Delta_C$.

Proof. The proof follows from Theorem 1 and the definition of $\mathsf{gather}^A_\odot$ and $\mathsf{gather}^D_\odot$. □

The next definition introduces functions $\mathsf{gather}^A_{\odot_a,\odot_d}$ and $\mathsf{gather}^D_{\odot_a,\odot_d}$, which follow the common conflict resolution strategy of giving precedence to activating and deactivating hyper-edges, respectively.

Definition 17. *Let M be an (A, Σ)-PRG and $\odot_a, \odot_d \in \{\Sqcap, \Sqcup\}$. The function*

$$
\begin{aligned}
\mathsf{gather}^A_{\odot_a,\odot_d}(e_s, e_t) \, : \, E \times E &\longrightarrow A^2 \\
(e_s, e_t) &\longmapsto \begin{cases} \mathsf{gather}^A_{\odot_a}(e_s, e_t) & \textit{if } \Gamma_A(e_s, e_t) \neq \emptyset \\ \mathsf{gather}^D_{\odot_d}(e_s, e_t) & \textit{otherwise} \end{cases}
\end{aligned}
$$

resolves conflicting hyper-edges by prioritizing activating hyper-edges when present. Similarly, the function

$$\mathsf{gather}^{D}_{\odot_a,\odot_d}(e_s,e_t) : E \times E \longrightarrow A^2$$

$$(e_s,e_t) \longmapsto \begin{cases} \mathsf{gather}^{D}_{\odot_a}(e_s,e_t) & \text{if } \Gamma_D(e_s,e_t) \neq \emptyset \\ \mathsf{gather}^{A}_{\odot_d}(e_s,e_t) & \text{otherwise} \end{cases}$$

resolves conflicts by prioritizing deactivating hyper-edges when present.

Intuitively, function $\mathsf{gather}^{A}_{\odot_a,\odot_d}(e_s,e_t)$ considers only the activating hyper-edges if any are present (i.e., $\Gamma_A(e_s,e_t) \neq \emptyset$), and $\mathsf{gather}^{D}_{\odot_a,\odot_d}(e_s,e_t)$ follows the opposite strategy, giving priority to deactivating hyper-edges.

As stated, we are interested in combining information from all hyper-edges when resolving conflicts. Hence, we introduce the function $\mathsf{gather}^{\odot}_{\odot_a,\odot_d}$, which, unlike the functions in Definition 17, does not prioritize one type of hyper-edges over the other. Thus, whenever there is a conflict between hyper-edges of different types, as illustrated in Figure 5, the function $\mathsf{gather}^{\odot}_{\odot_a,\odot_d}$ first combines the information from activating and deactivating hyper-edges separately, and then combines both results.

Definition 18. *Let M be an (A,Σ)-PRG and $\odot \in \{\sqcap\!\!\!\sqcap, \sqcup\!\!\!\sqcup, \otimes, \oplus\}$ and $\odot_a, \odot_d \in \{\sqcap\!\!\!\sqcap, \sqcup\!\!\!\sqcup\}$. The function $\mathsf{gather}^{\odot}_{\odot_a,\odot_d}(e_s,e_t) : E \times E \longrightarrow A^2$ combines the information from all conflicting hyper-edges that are triggered by e_s and target e_t as follows*

$$\mathsf{gather}^{\odot}_{\odot_a,\odot_d}(e_s,e_t) = \begin{cases} \mathsf{gather}^{A}_{\odot_a}(e_s,e_t) & \text{if } \Gamma_A(e_s,e_t) \neq \emptyset \text{ and } \Gamma_D(e_s,e_t) = \emptyset \\ \mathsf{gather}^{D}_{\odot_d}(e_s,e_t) & \text{if } \Gamma_A(e_s,e_t) = \emptyset \text{ and } \Gamma_D(e_s,e_t) \neq \emptyset \\ \mathsf{gather}^{A}_{\odot_a}(e_s,e_t) \odot \mathsf{gather}^{D}_{\odot_d}(e_s,e_t) & \text{otherwise} \end{cases}$$

Informally, if there are only activating hyper-edges, we combine their information using the function $\mathsf{gather}^{A}_{\odot_a}$. Similarly, if there are deactivating hyper-edges only, we use $\mathsf{gather}^{D}_{\odot_d}$. Finally, if there is a mixture of activating and deactivating hyper-edges in $\Gamma(e_s,e_t)$, we first combine each type separately using $\mathsf{gather}^{A}_{\odot_a}$ and $\mathsf{gather}^{D}_{\odot_d}$ respectively, and then combine the results using $\odot$.

This allows us to have sixteen ways to combine conflicts with function $\mathsf{gather}^{\odot}_{\odot_a,\odot_d}$ since $\odot_a, \odot_d \in \{\sqcap\!\!\!\sqcap, \sqcup\!\!\!\sqcup\}$ and $\odot \in \{\sqcap\!\!\!\sqcap, \sqcup\!\!\!\sqcup, \otimes, \oplus\}$. The different choices give rise to different policies that should align with the problem at hand.

The choice of operators $\odot_a$ and $\odot_d$ has been addressed before and it boils down to whether we value any available evidence or require consensus among hyper-edges for the activation or deactivation of an edge. Whereas the choice of $\odot$, which is relevant when $\Gamma_A(e_s,e_t) \neq \emptyset$ and $\Gamma_D(e_s,e_t) \neq \emptyset$, essentially carries the same intuition and meaning as the operator $\odot$ in the twist-structure.

That is, $\mathsf{gather}^{\sqcap}_{\odot_a,\odot_d}$ (resp. $\mathsf{gather}^{\sqcup}_{\odot_a,\odot_d}$) corresponds to the usual "and" (resp. "or"), so the function performs the conjunction (resp. disjunction) of the information given by the activating hyper-edges and by the deactivating hyper-edges, which corresponds again to the "worst-case" (resp. "best-case") scenario.

Function $\mathsf{gather}^{\otimes}_{\odot_a,\odot_d}$ boils down to a "consensus" between the information given by the activating and deactivating hyper-edges. Meanwhile, $\mathsf{gather}^{\oplus}_{\odot_a,\odot_d}$ stands for "accept all", i.e., it combines all the knowledge provided by the activating and deactivating hyper-edges.

The following example illustrates how information from conflicting hyper-edges is combined using the function $\mathsf{gather}^{\odot}_{\odot_a,\odot_d}$.

Example 16. *Consider the following* $(\{0,1\},\emptyset)$*-PRG*[6]*.*

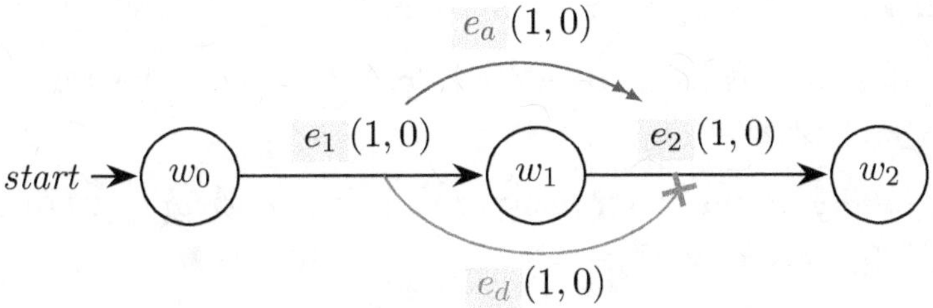

This is a scenario involving conflicting hyper-edges: according to e_a*, the target edge* e_2 *should be fully activated, while according to* e_d*, it should be fully deactivated. Formally,*

$$\begin{aligned}\mathsf{gather}^{A}_{\odot_a}(e_1,e_2) &= \mathsf{Act}(\mathcal{C}(e_1),\mathcal{C}(e_a),\mathcal{C}(e_2)) = (1,0)\\ \mathsf{gather}^{D}_{\odot_d}(e_1,e_2) &= \mathsf{Deact}(\mathcal{C}(e_1),\mathcal{C}(e_d),\mathcal{C}(e_2)) = (0,1)\end{aligned}$$

Let us see the result of combining the information from e_a *and* e_d *using the function* $\mathsf{gather}^{\odot}_{\odot_a,\odot_d}(e_s,e_t)$*, while varying the choice of operator* $\odot$*. Since the conflict involves only one activating and one deactivating hyper-edge, the specific choices of* $\odot_a$ *and* $\odot_d$ *are irrelevant.*

Therefore, following the traversal of e_1*, the value of* e_2 *is updated according to the choice of* $\odot$*. This results in the following possibilities:*

- $\mathsf{gather}^{\sqcap}_{\odot_a,\odot_d}(e_s,e_t) = (1,0) \sqcap (0,1) = (0,1)$ *which yields the most conservative or "worst-case" scenario where* e_2 *is completely deactivated.*

[6] Note that since all transitions are classically true, i.e. $(1,0)$, this PRG could be represented as a standard reactive graph.

- $\mathsf{gather}^{\Cup}_{\odot_a,\odot_d}(e_s, e_t) = (1,0) \Cup (0,1) = (1,0)$ *which yields the most optimistic or "best-case" scenario where e_2 is completely activated.*
- $\mathsf{gather}^{\otimes}_{\odot_a,\odot_d}(e_s, e_t) = (1,0) \otimes (0,1) = (1 \sqcap 0,\ 0 \sqcap 1) = (0,0)$ *represents the "consensus" between all conflicting hyper-edges. Since e_a and e_d disagree completely, the result reflects complete vagueness/ignorance.*
- $\mathsf{gather}^{\oplus}_{\odot_a,\odot_d}(e_s, e_t) = (1,0) \oplus (0,1) = (1 \sqcup 0,\ 0 \sqcup 1) = (1,1)$ *represents "accepting all" information given by the conflicting hyper-edges, which yields a complete contradiction, the system accepts both the activation and deactivation.*

The following corollary follows from Corollary 1 and Theorems 1 and 2.

Corollary 2. *Let M be any (A,Σ)-PRG and $\odot_a, \odot_d \in \{\Cup, \Cap\}$.*

- *If M is totally consistent, then for $e_s, e_t \in E$ and $\odot \in \{\Cap, \Cup, \otimes\}$ it follows that* $\mathsf{gather}^{\odot}_{\odot_a,\odot_d}(e_1, e_2) \in \Delta_C$.
- *If M is totally inconsistent, then for $e_s, e_t \in E$ and for $\odot \in \{\Cap, \Cup, \oplus\}$ it follows that* $\mathsf{gather}^{\odot}_{\odot_a,\odot_d}(e_1, e_2) \in \Delta_P$.

Note that the operator $\oplus$ does not preserve consistency, and $\otimes$ does not preserve knowledge. Hence, choosing to "accept all" information may result in inconsistencies, while opting for "gullibility" may lead to a loss in the overall amount of knowledge.

We now aim to define a semantics by "unfolding" any PRG into a paraconsistent graph. As before, we first describe how the activation function $\mathcal{C}$ of a PRG is updated after traversing an edge, using the conflict resolution functions introduced in Definitions 17 and 18.

Definition 19. *Let M be an (A,Σ)-PRG, and let $\mathcal{C}$ its current activation function. Let $\odot_a, \odot_d \in \{\Cap, \Cup\}$ and $\odot \in \{\Cap, \Cup, \otimes, \oplus\}$ be operators defining the conflict resolution policies. For any edges $e, x \in E$, we define the updated value of edge x after traversing edge e as follows:*

$$\mathcal{C}^{e}_{A,\odot_a,\odot_d}(x) = \begin{cases} \mathcal{C}(x) & \text{if } x \notin (\mathsf{on}(e) \cap \mathsf{off}(e)) \\ \mathsf{gather}^{A}_{\odot_a,\odot_d}(e, x) & \text{otherwise} \end{cases} \tag{17}$$

when conflicts are resolved using the function $\mathsf{gather}^{A}_{\odot_a,\odot_d}$.

$$\mathcal{C}^{e}_{D,\odot_a,\odot_d}(x) = \begin{cases} \mathcal{C}(x) & \text{if } x \notin (\mathsf{on}(e) \cap \mathsf{off}(e)) \\ \mathsf{gather}^{D}_{\odot_a,\odot_d}(e, x) & \text{otherwise} \end{cases} \tag{18}$$

when conflicts are resolved using the function $\mathsf{gather}^{D}_{\odot_a,\odot_d}$.

$$\mathcal{C}^{e}_{\odot,\odot_a,\odot_d}(x) = \begin{cases} \mathcal{C}(x) & \text{if } x \notin (\mathsf{on}(e) \cap \mathsf{off}(e)) \\ \mathsf{gather}^{\odot}_{\odot_a,\odot_d}(e,x) & \text{otherwise} \end{cases} \tag{19}$$

when conflicts are resolved using the function $\mathsf{gather}^{\odot}_{\odot_a,\odot_d}$.

In other words, if edge x is not targeted by any hyper-edge triggered by e, its value remains unchanged. Otherwise, the information from all hyper-edges triggered by e and targeting x is combined using one of the functions: $\mathsf{gather}^{A}_{\odot_a,\odot_d}$, $\mathsf{gather}^{D}_{\odot_a,\odot_d}$, or $\mathsf{gather}^{\odot}_{\odot_a,\odot_d}$, depending on the chosen policies.

We are now equipped to define the semantics for PRG with possibly conflicting hyper-edges. As before, we begin from an initial configuration $\langle w_0, \mathcal{C}_0\rangle$. Then, for any ground edge $e \in \twoheadrightarrow$ originating from the initial state w_0, we compute the updated activation function for all edges x using the update functions defined in Definition 19, according to the chosen conflict resolution policy.

Definition 20. *The semantics of an* (A, Σ)*-paraconsistent graph* P *is defined as the evolution of a configuration* $\langle w, \mathcal{C}\rangle$*, where* $w \in W$ *and* $\mathcal{C} : E \to A^2$*, starting from* $\langle w_0, \mathcal{C}_0\rangle$*. Depending on the conflict resolution policy employed, the system evolves according to the following rule:*

- *If using* $\mathsf{gather}^{A}_{\odot_a,\odot_d}$*:*

$$\frac{\exists e \in \twoheadrightarrow \cdot\ \overline{e} = (w, a, w')}{\langle w, \mathcal{C}\rangle \xrightarrow{a,\mathcal{C}(e)}_P \langle w', \mathcal{C}^{e}_{A,\odot_a,\odot_d}\rangle}$$

- *If using* $\mathsf{gather}^{D}_{\odot_a,\odot_d}$*:*

$$\frac{\exists e \in \twoheadrightarrow \cdot\ \overline{e} = (w, a, w')}{\langle w, \mathcal{C}\rangle \xrightarrow{a,\mathcal{C}(e)}_P \langle w', \mathcal{C}^{e}_{D,\odot_a,\odot_d}\rangle}$$

- *If using* $\mathsf{gather}^{\odot}_{\odot_a,\odot_d}$*:*

$$\frac{\exists e \in \twoheadrightarrow \cdot\ \overline{e} = (w, a, w')}{\langle w, \mathcal{C}\rangle \xrightarrow{a,\mathcal{C}(e)}_P \langle w', \mathcal{C}^{e}_{\odot,\odot_a,\odot_d}\rangle}$$

where $\odot_a, \odot_d \in \{\sqcup\!\!\sqcup, \sqcap\!\!\sqcap\}$ *and* $\odot \in \{\sqcap\!\!\sqcap, \sqcup\!\!\sqcup, \otimes, \oplus\}$.

If the paraconsistent reactive graph does not contain conflicting hyper-edges, then the conflict resolution policies introduced through the functions $\mathsf{gather}^{A}_{\odot_a,\odot_d}$, $\mathsf{gather}^{D}_{\odot_a,\odot_d}$, and $\mathsf{gather}^{\odot}_{\odot_a,\odot_d}$ become unnecessary. Consequently, all of the rules in Definition 20 reduce to the semantics for PRG without conflicting hyper-edges, as given in Definition 15, as expected.

Furthermore, the following result is immediate from Corollary 2.

Corollary 3. *Let M be an (A, Σ)-PRG.*

- *If M is totally consistent, then the system evolves without ever generating inconsistency under any of the rules in Definition 20, except when using the resolution function* $\mathsf{gather}^{\oplus}_{\odot_a,\odot_d}$.
- *If M is totally inconsistent, then the system evolves without ever becoming consistent under any of the rules in Definition 20, except when using the resolution function* $\mathsf{gather}^{\otimes}_{\odot_a,\odot_d}$.

Intuitively, the above corollary clarifies that some conflict-resolution policies preserve inconsistency while others preserve consistency. As shown in Theorem 1, operator $\oplus$ does not preserve consistency, whereas Theorem 2 shows that $\otimes$ does not preserve inconsistencies. Consequently, when these operators are used to resolve conflicts between hyper-edges, via $\mathsf{gather}^{\oplus}_{\odot_a,\odot_d}$ and $\mathsf{gather}^{\otimes}_{\odot_a,\odot_d}$, the unfolding rules of Definition 20 proceed in a way that may affect whether the resulting system will remain consistent or inconsistent.

For instance, as illustrated in Example 16, applying $\mathsf{gather}^{\oplus}_{\odot_a,\odot_d}$ (where $\oplus$ is informally described as "accept all") may introduce inconsistencies even when all transitions initially carry consistent values. Likewise, using $\mathsf{gather}^{\otimes}_{\odot_a,\odot_d}$ (with $\otimes$ informally understood as a "consensus" operator) may fail to preserve inconsistent information; that is, transitions that are originally inconsistent may be updated to consistent ones. These behaviours account for the exceptions stated in Corollary 3.

5 Conclusions

This work is part of a broader research agenda on paraconsistent structures [16, 18, 19]. The conference paper it extends [17] first combined this agenda with reactive graphs [27] by introducing reactive graphs with paraconsistent transitions. These allow ground edges to represent consistent, vague, or inconsistent information. The main contribution was the definition of a semantics that "unfolds" an $\mathrm{RG_P}$ into a paraconsistent graph by eliminating hyper-edges while unfolding the system's reactivity.

Introducing paraconsistency into reactive graphs provides a mechanism for modeling evolving systems where information may be incomplete, contradictory, or subject to revision as new evidence emerges. Such scenarios frequently arise in domains involving multiple agents or sources, where beliefs, lies, uncertainty, or disagreement must be reconciled over time.

This paper generalizes reactive graphs with Paraconsistent Transition by introducing paraconsistent reactive graphs, abbreviated as PRG. In PRG, all edges, both ground and hyper-edges, can represent vague, consistent, or inconsistent information. This generalization is motivated by the observation that, if hyper-edges are interpreted as information provided by different sources, then conflicts and inconsistencies naturally arise. Initially, in Section 4.1, we restricted our focus to PRG without conflicting hyper-edges in order to study the isolated effects of activation and deactivation. In Section 4.2, we addressed however the more general setting where multiple hyper-edges may simultaneously target the same edge, potentially conveying contradictory information.

To manage such conflicts, we proposed a range of resolution strategies, ranging from conservative to optimistic, that allow for flexible combinations of information. In contrast to bivalent switch graphs, where conflict resolution typically defaults to a binary choice (e.g., prioritizing activation over deactivation [30, 42]), our framework accommodates more nuanced policies for handling conflicts. Each resolution strategy is grounded in an intuitive conflict policy that should be chosen according to the problem at hand.

Future work will explore refined conflict resolution policies that go beyond simply combining all conflicting hyper-edges. For instance, one may prioritize hyper-edges based on their degree of consistency or informativeness. A possible approach is to rank hyper-edges by how much information they hold and select the most informative or least inconsistent ones. Another option is to assign trust scores (e.g., values in $[0, 1]$) to hyper-edges, reflecting prior knowledge or confidence in their reliability. These enhancements would require revisiting the definitions of the Act and Deact functions (see Example 12), as well as the gather functions introduced in Section 4.2.

Additional ongoing research includes addressing the challenges posed by various application scenarios of paraconsistent structures, spanning diverse fields from robotics to quantum computation [4].

References

[1] Ofer Arieli and Arnon Avron. Reasoning with logical bilattices. *J. Log. Lang. Inf.*, 5(1):25–63, 1996.

[2] Ofer Arieli, Chris Cornelis, Glad Deschrijver, and Etienne Kerre. Bilattice-based squares and triangles. In Lluís Godo, editor, *Symbolic and Quantitative Approaches to Reasoning with Uncertainty*, pages 563–575, Berlin, Heidelberg, 2005. Springer Berlin Heidelberg.

[3] Florencio G. Asenjo. A calculus of antinomies. *Notre Dame J. Formal Log.*, 7(1):103–105, 1966.

[4] Luís Soares Barbosa and Alexandre Madeira. Capturing qubit decoherence through paraconsistent transition systems. In Shigeru Chiba, Youyou Cong, and Elisa Gonzalez Boix, editors, *Companion Proceedings of the 7th International Conference on the Art, Science, and Engineering of Programming, Programming 2023, Tokyo, Japan, March 13-17, 2023*, pages 109–110. ACM, 2023.

[5] Nuel D. Belnap. *A Useful Four-Valued Logic*, pages 5–37. Springer Netherlands, Dordrecht, 1977.

[6] Félix Bou, Francesc Esteva, Lluís Godo, and Ricardo Oscar Rodríguez. On the Minimum Many-Valued Modal Logic over a Finite Residuated Lattice. *Journal of Logic and Computation*, 21(5):739–790, 10 2009.

[7] Suene Campos, Daniel Figueiredo, Manuel A. Martins, and Regivan Santiago. Labeled fuzzy reactive graphs. *Fuzzy Sets and Systems*, 510:109320, 2025.

[8] Suene Campos, Regivan Santiago, Manuel A. Martins, and Daniel Figueiredo. Introduction to reversal fuzzy switch graph. *Science of Computer Programming*, 216:102776, 2022.

[9] Suene Campos, Regivan Santiago, Manuel A. Martins, and Daniel Figueiredo. Aggregation-based operations for reversal fuzzy switch graphs. *Fuzzy Sets and Systems*, 466:108273, 2023.

[10] Walter Carnielli and Marcelo Esteban Coniglio. *Paraconsistent Logic: Consistency, Contradiction and Negation.* Springer International Publishing, 2016.

[11] Diana Costa, Daniel Figueiredo, and Manuel A. Martins. Relation-changing models meet paraconsistency. *Journal of Logical and Algebraic Methods in Programming*, 133:100870, 2023.

[12] Diana Costa and Manuel A. Martins. Paraconsistency in hybrid logic. *Journal of Logic and Computation*, 27(6):1825–1852, 10 2016.

[13] Diana Costa and Manuel A. Martins. Measuring inconsistent diagnoses. In *2018 IEEE 20th International Conference on e-Health Networking, Applications and Services (Healthcom)*, pages 1–4, 2018.

[14] Ana Cruz, Alexandre Madeira, and Luís Soares Barbosa. A logic for paraconsistent transition systems. In Andrzej Indrzejczak and Michal Zawidzki, editors, *10th International Conference on Non-Classical Logics. Theory and Applications*, volume 358 of *EPTCS*, pages 270–284, 2022.

[15] Ana Cruz, Alexandre Madeira, and Luís Soares Barbosa. Paraconsistent transition systems. In Daniele Nantes-Sobrinho and Pascal Fontaine, editors, *Proceedings 17th International Workshop on Logical and Semantic Frameworks with Applications, LSFA*

2022, Belo Horizonte, Brazil (hybrid), 23-24 September 2022, volume 376 of *EPTCS*, pages 3–15, 2022.

[16] Juliana Cunha, Alexandre Madeira, and Luís Soares Barbosa. Stepwise development of paraconsistent processes. In Cristina David and Meng Sun, editors, *Theoretical Aspects of Software Engineering - 17th International Symposium, TASE 2023, Bristol, UK, July 4-6, 2023, Proceedings*, volume 13931 of *Lecture Notes in Computer Science*, pages 327–343. Springer, 2023.

[17] Juliana Cunha, Alexandre Madeira, and Luís Soares Barbosa. Paraconsistent reactive graphs. In , editor, *The International Workshop on Reconfigurable Transition Systems: Semantics, Logics and Applications, ReacTS 2024, Aveiro, Portugal, November 4-5, 2024, Proceedings*, volume (in print) of *Lecture Notes in Computer Science*. Springer, 2024.

[18] Juliana Cunha, Alexandre Madeira, and Luís Soares Barbosa. Specification of paraconsistent transition systems, revisited. *Science of Computer Programming*, 240:103196, 2025.

[19] Juliana Cunha, Alexandre Madeira, and Luís S. Barbosa. Paraconsistent relations as a variant of Kleene algebras. *Electronic Proceedings in Theoretical Computer Science*, 421:130–147, June 2025.

[20] Newton C. A. da Costa. On the theory of inconsistent formal systems. *Notre Dame J. Formal Log.*, 15(4):497–510, 1974.

[21] Mustafa Demirci. Aggregation operators on partially ordered sets and their categorical foundations. *Kybernetika*, 42(3):261–277, 2006.

[22] Glad Deschrijver, Ofer Arieli, Chris Cornelis, and Etienne E. Kerre. A bilattice-based framework for handling graded truth and imprecision. *Int. J. Uncertain. Fuzziness Knowl. Based Syst.*, 15(1):13–41, 2007.

[23] M. Fitting. Bilattices in logic programming. In *Proceedings of the Twentieth International Symposium on Multiple-Valued Logic*, pages 238–246, 1990.

[24] Melvin Fitting. Bilattices and the semantics of logic programming. *The Journal of Logic Programming*, 11(2):91–116, 1991.

[25] D. M. Gabbay, J.-Y. Beziau, and W. Carnielli, editors. *Handbook of Paraconsistency*. College Publications, 2007.

[26] Dov Gabbay and Sérgio Marcelino. Global view on reactivity: switch graphs and their logics. *Annals of Mathematics and Artificial Intelligence*, 66(1):131–162, Dec 2012.

[27] Dov M. Gabbay. *Introducing Reactive Kripke Semantics and Arc Accessibility*, pages 292–341. Springer Berlin Heidelberg, Berlin, Heidelberg, 2008.

[28] Dov M. Gabbay. Reactive kripke models and contrary to duty obligations. part A: semantics. *J. Appl. Log.*, 11(1):103–136, 2013.

[29] Dov M. Gabbay. *Reactive Kripke Semantics*. Cognitive Technologies. Springer, 2013.

[30] Dov M. Gabbay and Sérgio Marcelino. Modal logics of reactive frames. *Stud Logica*, 93(2-3):405–446, 2009.

[31] Matthew L. Ginsberg. Multivalued logics: a uniform approach to reasoning in artificial

intelligence. *Comput. Intell.*, 4:265–316, 1988.

[32] Michel Grabisch, Jean-Luc Marichal, Radko Mesiar, and Endre Pap. *Aggregation Functions.* Encyclopedia of Mathematics and its Applications. Cambridge University Press, 2009.

[33] Petr Hájek. Basic fuzzy logic and bl-algebras. *Soft Comput.*, 2(3):124–128, 1998.

[34] Stanisław Jaśkowski. A propositional calculus for inconsistent deductive systems. *Logic and Logical Philosophy*, 7(7):35–56, Jan. 2004.

[35] J. A. Kalman. Lattices with involution. *Transactions of the American Mathematical Society*, 87:485–491, 1958.

[36] Magda Komorníková and Radko Mesiar. Aggregation functions on bounded partially ordered sets and their classification. *Fuzzy Sets and Systems*, 175(1):48–56, 2011. Theme: Aggregation Functions, Generalised Measure Theory.

[37] Marcus Kracht. On extensions of intermediate logics by strong negation. *Journal of Philosophical Logic*, 27(1):49–73, 1998.

[38] Orna Kupferman and Yoad Lustig. Lattice automata. In Byron Cook and Andreas Podelski, editors, *Verification, Model Checking, and Abstract Interpretation*, pages 199–213, Berlin, Heidelberg, 2007. Springer Berlin Heidelberg.

[39] Mohammad Reza Mousavi and Mahsa Varshosaz. Telling lies in process algebra. In Jun Pang, Chenyi Zhang, Jifeng He, and Jian Weng, editors, *2018 International Symposium on Theoretical Aspects of Software Engineering, TASE 2018, Guangzhou, China, August 29-31, 2018*, pages 116–123. IEEE Computer Society, 2018.

[40] Graham Priest. Paraconsistency and dialetheism. In Dov M. Gabbay and John Woods, editors, *The Many Valued and Nonmonotonic Turn in Logic*, volume 8 of *Handbook of the History of Logic*, pages 129–204. Elsevier, 2007.

[41] Regivan Santiago, Manuel A. Martins, and Daniel Figueiredo. Introducing fuzzy reactive graphs: a simple application on biology. *Soft Computing*, 25(9):6759–6774, May 2021.

[42] David Tinoco, Alexandre Madeira, Manuel A. Martins, and José Proença. *Reactive Graphs in Action*, page 97–105. Springer Nature Switzerland, 2024.

Received June 2025

Beyond the Spell: A Dynamic Logic Analysis of Misdirection

Benjamin Icard
LIP6, Sorbonne University, CNRS, France

Raul Fervari
Consejo Nacional de Investigaciones Científicas y Técnicas (CONICET) and Universidad Nacional de Córdoba, FAMAF, Argentina

Abstract

Misdirection can be defined as the intentional action of causing some misrepresentation in an agent, or in a group of agents. Such misrepresentations may result from verbal actions, as in linguistic deception, or from visual actions, as in visual misdirection. Examples of visual misdirection abound (e.g. in nature, in the military), with magic tricks providing a vivid illustration. So far, various types of verbal misdirection have been investigated from a formal perspective (e.g. lying, bluffing) but little attention has been paid to the particular case of visual misdirection. In this paper, we introduce a dynamic epistemic logic to represent not only verbal misdirection on agents' beliefs but also visual misdirection on agents' observations. We illustrate the dynamics of the logic by modelling a classic magic trick known as the French Drop. We also provide a sound and complete axiom system for the logic, and discuss the expressivity and scope of the setting.

1 Introduction

Human misdirection refers to an agent's intentional action to make addressees form inaccurate representations of the world by "*deflecting [their] attention for the purpose of disguise*" [54]. These inaccurate representations, or *misrepresentations*, can arise through different modalities, either strictly verbal (e.g., announcements) or non-verbal (e.g., visual cues) [6, 35]. When the modality is verbal, as in announcements, misrepresentations usually consist in inaccurate beliefs (with the notion of belief defined as in [65]), that can be associated with observations, as in so-called "*epistemic seeing*" [20, 21]. Therein, it is also established that when the modality is non-verbal, as in visual observation, misrepresentations consist in inaccurate observations that can involve inaccurate beliefs (i.e., inaccurate *epistemic seeing*) but not necessarily, as in so-called "*non-epistemic seeing*". To the best of our knowledge, no formal treatment has been given to visual misdirection and resulting inaccurate observations, in contrast to verbal misdirection. To fill this gap, we propose to use dynamic epistemic logic (DEL) to analyze both verbal and visual misdirection, their similarities and differences, and resulting misrepresentations. For clarity, we call "strategies of misdirection" the pairings of actions with the misrepresentations they are designed to produce.

In the conceptual and logical literature, misdirection has been mainly studied through linguistic lenses, i.e. as verbal deception. As the prototypical form of verbal deception [15, 41], lying has received the most analytical attention (e.g., [51, 63]). However, related borderline strategies, such as *bluffing* [48, 62], *half-truths* [50], and *omission of information* [50, 52], have also been formally analyzed. In the course of those investigations, visual strategies of misdirection have been largely overlooked, – even though the concept of visual observation has caught some attention [16, 31].

Visual misdirection is pervasive in humans [6, 29] as well as in nature [46, 47]. A fundamental distinction concerning misdirection is between *dissimulation* and *simulation*. In [6], dissimulation is defined as the action of hiding facts to deceive addressees, whereas simulation consists in deceiving them by showing some unreal facts as if they were really happening. Important to notice is that, in nature, camouflage and faking embody this broader distinction between *dissimulation* and *simulation* in misdirection. In *camouflage*, one individual aims to blend into its environment to remain undetected from predators, while *faking* consists of simulating a false or counterfactual reality to distract their attention. Dissimulation and simulation actually occur in all living entities. In humans, those strategies have been practiced since ancient times by the military, e.g., *simulation* with the Trojan horse [69]. But, actually, stage magicians have also played a central role, even in the military domain [42] by crafting visual illusions to trick enemies, based on *simulation* (with fake tanks) or on *dissimulation* (with light effects) [27].

Main contributions. This paper aims to deepen the understanding of misdirection through a logical analysis of its verbal and visual forms. Rather than proposing a new theory of misdirection, we formalize established conceptual distinctions, including the contrast between non-epistemic and epistemic seeing [20, 21] and that between visual simulation and dissimulation [6]. Our goal is to clarify, from a logical perspective, how these distinctions interact and what role each plays in the production of misdirection. To this end, we analyze the French Drop, a classical magic trick (see, e.g. [5, 35]), in which a performer appears to transfer a coin from one hand to the other while both simulating the transfer and dissimulating the coin. Logical work on magic has already shown how formal tools can illuminate magical effects, for instance by analyzing the sense of impossibility they produce [57]. We likewise address the cognitive effect of surprise in the French Drop, but our main objective is to characterize the deceptive structure of misdirection, verbal and visual, in magic and beyond. We consider this a starting point in modeling different forms of misdirection and their properties.

To carry out this analysis, we introduce a dynamic logic of misdirection, **DLM**, for truthful and deceptive verbal and visual actions. Technically, **DLM** adapts standard dynamic epistemic logic [2, 64] by combining atoms for visual observation, a doxastic modality for belief, and a dynamic modality for updating beliefs and observations under verbal announcements and visual actions. Our treatment of visual observation is inspired by the logic of (in)attentive agents [4] (see also [11]), but departs from it in two respects: observing p neither amounts to attending to p nor entails believing p, and we distinguish syntactically between observing p and observing $\neg p$. As in [4], not observing a fact does not imply observing its negation, so the relevant models must satisfy suitable constraints. The dynamic component relies on action models with postconditions [61, 60], including models in which one designated agent misdirects the others. This framework allows us to track, step by step, how actions affect belief and visual observation. It also lets us formalize not only the French Drop, with emphasis on visual simulation, but also related notions such as dissimulation, stronger notions of belief and observation, and surprise as a cognitive effect. Moreover, even though the original logical machinery we introduce is rather simple, we argue that this work still constitutes a novel, interesting application of the Dynamic Epistemic Logic family.

Outline of the paper. In Section 2, we discuss the concepts which motivate our proposal and technical approach. We review the logical literature on the formalization of misdirection, and put emphasis on missing features in extant proposals. Section 3 introduces our dynamic logic of misdirection called **DLM**. This section also provides a sound and complete axiom system for **DLM** over the appropriate class of models. In Section 4, we show that a small fragment of the logic called $\mathsf{DLM}^{(+,-)}$ is sufficient for our expressive purposes, since dynamic modalities only make use of four (types of) action models to capture genuine and deceptive actions, either verbal or visual. Section 4.2 illustrates the relevance of **DLM** by modelling the dynamics of the French Drop trick. In Section 5, we show that our setting can express stronger belief attitudes, epistemic versions of observation, more complex notions of misdirection and also the sense of surprise resulting from magic misdirection. Section 6 provides some final remarks and future directions of research.

2 Setting the Stage for Misdirection

We discuss here key distinctions concerning agents' beliefs and observations. Then, we review extant logical frameworks for formalizing misdirection against the background of the classical distinction between *dissimulation* and *simulation.* Finally, we identify the logical features required to capture verbal and visual misdirection.

2.1 Believing and Observing

For simplicity, we distinguish two perspectives, that of the operator, who acts on the basis of some information, and that of the addressee, who reacts to the information received. Among possible modalities, we focus on verbal and visual actions, setting aside others such as auditory, kinesthetic, or olfactory ones. A verbal case is that of a politician addressing potential voters so as to make them believe that he or she is the best candidate. A visual case is that of a stage magician performing for spectators so as to make them observe an unreal event, such as the appearance or disappearance of a rabbit. In verbal contexts, the operator and addressee are usually called the "speaker" and the "hearer". In visual contexts, the operator may be called the "performer", and the addressees the "spectators", "audience", or "public".

Let us start by drawing distinctions concerning verbal and visual attitudes before considering the actions affecting those attitudes. Agents can entertain three possible, yet non exclusive, attitudes with respect to their surroundings:

- ***Observing with believing***, called here "epistemic observation", corresponds to Dretske's notion of "epistemic seeing" [20, 21]. Some objects, such as an apple or a parrot, are not merely seen as shapes but seen *as* an apple or *as* a parrot. Likewise, a detective may observe pieces of evidence not as scattered objects and marks, but as a crime scene indicating a struggle or a crime. The same applies to facts or events. An agent epistemically sees a car moving forward in the street when she sees it *as* a car moving forward, rather than as an unidentified moving object. In such cases, observation involves conscious recognition and yields explicit beliefs about what is observed.

- ***Believing without observing***, called herein "simple belief", concerns cases in which agents believe that certain events occur beyond their sight without observing them directly. For example, an agent may believe that a car has crashed on the street on the basis of verbal information from a direct observer, even though she has not witnessed the event herself.

  Observing without believing, called herein "atomic observation", corresponds to Dretske's notion of "simple seeing" [20, 21] and to "non-epistemic seeing" in [56, 19]. It can be understood as a causal process. When an object, such as an apple or a parrot, is in an agent's field of vision and light reaches the eyes, it may be perceived without being recognized *as* an apple or *as* a parrot. Likewise, a person looking through a window may observe the street in this atomic, that is, non-epistemic, way if she forms no explicit doxastic representation of what occurs there, for instance of a parrot flying or a car moving forward, because she remains absorbed in her thoughts.

To sum up, belief necessarily involves a doxastic stance toward the world, whereas observation does not [20, 21]. In *epistemic* observation, the agent does adopt such a stance, since she consciously recognizes what she observes and thereby forms beliefs about it. In *atomic*, i.e., *non-epistemic*, observation, by contrast, the agent forms no belief about what she sees, as when she does not attend to what is happening in her immediate surroundings.

Regarding actions, verbal modalities involve acts of *telling* facts to addressees, whereas visual modalities involve acts of *showing* them. In both cases, successful actions produce epistemic effects in addressees, in different ways. Showing leads addressees to observe facts or events *epistemically*, not merely to see them. The act of showing directs their attention to what is presented, and thereby prompts explicit beliefs about it.Telling, by contrast, leads either to simple belief, when addressees cannot observe the relevant facts themselves, or to epistemic observation, when they can. In both cases, the addressee acquires new beliefs that change her uncertainty about the world. These distinctions are summarized in Figure 1.

Operators also separate into two categories regarding their intentions to addressees: *genuine* versus *misleading.* Intentions are genuine when the operators' goal is to inform addressees, while their intentions are misleading when they are deceptive and plan to mislead addressees. This is the case of misdirection. From the perspective of actions, misdirection can be verbal or visual. Verbal misdirection consists in telling addressees something false, whereas visual misdirection consists in presenting them with a false representation of reality.

2.2 Two Main Strategies of Misdirection

Building on [6], we formalize two central strategies of misdirection: *dissimulation of the truth* and *simulation of the false.* We treat them in their most general form, distinguishing only between their visual and verbal manifestations. For completeness, this section also reviews extant work on their main subcategories, to indicate the range of cases covered by our formal approach.

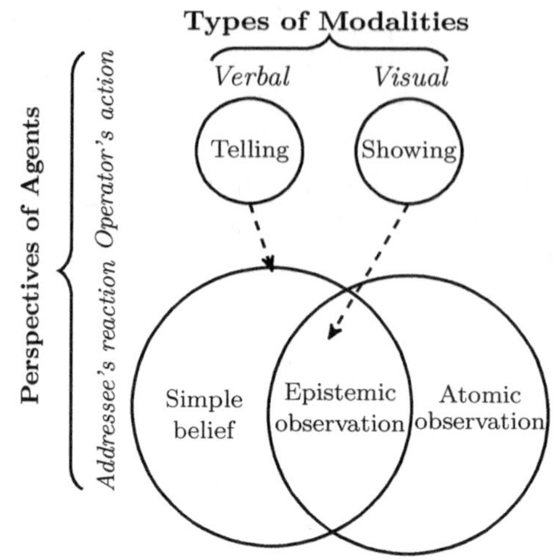

Figure 1: Possible interactions between agent's perspectives, or attitudes, regarding verbal versus visual informational modalities.

2.2.1 Misdirection as Dissimulation of the Truth

Dissimulation consists in hiding the truth to addressees to prevent them from making an accurate representation of the world. For e.g. [9, 25], hiding, or concealing information, consists in *masking* true information, — aiming addressees not to form accurate beliefs on the world. Important to notice is that dissimulation in this sense (i.e., hiding or concealing), differs from dissimulation as *withholding information*, or omission. The work in [15] insists on the fact that withholding information does not involve any effective action to deceive, contrary to hiding information which requires an effective action of concealment in that respect.

The effect of successful dissimulation is always *epistemic*: the addressee fails to reach accurate belief and/or accurate epistemic observation on her surroundings. In the verbal case, successful dissimulation leads addressees *not* to reach true beliefs, and can also lead them *not* to reach accurate epistemic observation when speakers use verbal omission to avoid attracting the visual attention of addressees to some specific fact or event. In the visual case now, successful dissimulation necessarily leads addressees to fail to epistemically observe some true reality. This is the case in the French Drop trick, where dissimulation is visual and relies on *masking*. As part of the trick, the magician conceals the coin by *palming* it into her left hand. In doing so, she puts a physical obstacle, namely her left hand, between the eyes of the public and the coin, thus concealing it from their sight. This makes spectators fail to epistemically observe that the coin is (still) in the magician's left hand. We argue that in the French Drop, this dissimulation through palming occurs simultaneously with the action of simulating a fake pass, which enhances the magical effect by diverting the audience's attention.

Verbal dissimulation has been formalized from two different perspectives. The first one has focused on expressing the *epistemic preconditions* of dissimulation [48, 63, 62]. By default, the precondition for dissimulation is the existence of a mismatch between the level of uncertainty the speaker has about the world, and the level of uncertainty he or she wants to convey to addressees. A variant on dissimulation is bluffing [48, 62]: now, speakers are completely uncertain about the truth or falsity of the utterances they make. But, as a precondition, the strongest form of uncertainty consists in full ignorance: in this case, speakers do not know, or at least believe to be true, the content they are uttering. Epistemic logicians have modelled different forms of ignorance but not from the sole perspective of deception [59, 26, 34].

The second perspective for modeling verbal dissimulation is more concerned with the *epistemic goals* of the deceivers. Following [18], [50] characterizes deception by *commission* in which deceivers perform some action to hide the truth, is distinguished from deception by *omission* in which deceivers do not perform any action to this end. Omission can be either *complete* or only *partial.* When it is complete, deceivers intend to hide the whole truth to addressees and thus, to prevent the formation of *any* belief in them. In this regard, [50, 52] have formalized the epistemic goals of complete omission under the name of *"withholding information"*. This strategy should be contrasted with the utterance of *"half-truths"* in which dissimulation is only partial. Following [23, 22], half-truths are vague utterances whose status is deliberately uncertain between truth and falsity, and that speakers may use to make addressees draw mistaken interpretations and conclusions. The epistemic goal of such half-truths is to keep addressees in the dark, as it is modeled in e.g. [52].

By contrast with verbal dissimulation, visual dissimulation has not been logically investigated so far, though the notions of observation and perception have caught logical attention in e.g. [31, 16, 3]. In visual dissimulation, one attempts to escape the gaze of others based on using various means, detailed in the taxonomy of deception proposed by [6]. One of those means, called *"masking"*, consists in preventing any potential observation by putting an obstacle between oneself (or the object to hide) and the observer in order to remain invisible (or the object). Another mean used for visual dissimulation is *"camouflage"*, or *"crypsis"*, in which the deceiver blends into its surroundings by adequately matching its colours and/or shapes. To the best of our knowledge, none of these distinctions has been formalized so far.

2.2.2 Misdirection as Simulation of the False

In contrast with dissimulation, simulation consists in presenting false information to addressees, either by using available false information or by fabricating false information, as in *faking* or *fabrication* [24]. The goal of simulation is to entertain addressees with inaccurate representations of the world.

As for dissimulation, the effect of successful simulation is always *epistemic* too, consisting in wrong beliefs and/or wrong epistemic observations for addressees. In the verbal case, successful simulation leads addressees to false beliefs, but can also lead them to false epistemic observation when addressees also see the physical manifestation of the simulated action. In the visual case now, successful simulation converts atomic observation into false epistemic observations for the addressees. This is the case in the French Drop trick: before the magician performs any visual action, the spectators observe the stage atomically since they do not necessarily pay attention to any specific object or fact of the scene. They just expect the magic trick to begin. Then, the magician enters the scene and the trick starts: she dissimulates the coin by palming into her left hand and then simulates the false action of moving the coin from left to right hand. By the magician performing those visual actions, the spectators' atomic observations become epistemic observations. At the end of the effect, indeed, the audience believe and epistemically observe that the coin has moved to the magician's right hand when this is not the case.

Extant formal literature on verbal simulation has given particular attention to the act of *lying*, as a prototypical instance of verbal simulation in which speakers utter propositional contents they believe to be false with the intention to deceive addressees. Intention (to deceive) has been formalized in static epistemic logic with classical Kripke models [51, 52], while their assertoric and deceptive dimensions have been expressed in dynamic epistemic logic [37, 63, 62, 1]. Using announcement operators to express intentionally false utterances, as in [37], [63, 62] logically distinguish *"subjective lies"* (i.e. believed to be false utterances that happen to be true in reality) from more *"objective lies"* (i.e. believed to be false utterances that also turn out to be really false). In [1], more exotic cases of lying are formalized, such as Moorean lies, i.e. utterances that are initially false but become true once they are announced. Apart from lies, verbal simulation can also consist in making addressees perform *wrong inferences* based on the speakers' utterances. In particular, [51] formalizes *"deductive lies"* in which speakers want addressees to infer inaccurate conclusions from their utterances, distinguishing them from *"abductive lies"* in which speakers expect them to jump directly to false conclusions, without any deduction [53].

Here again, by contrast with verbal simulation, visual simulation has not been logically investigated to the best of our knowledge. As mentioned, only the notions of observation and perception underlying visual simulation have caught logical attention recently [31, 16, 3]. In turn, our interest is to model these two concepts in general via dynamic actions changing the perception of an agent, incorporating also the notion of observation into the picture as a novelty.

2.3 Modelling Verbal and Visual Misdirection

2.3.1 Misdirection in the French Drop Trick

Misdirection has been analyzed extensively by stage magic practitioners (see e.g. [43, 28]). When magicians aim to master a new trick, they often break it down into two key components: the *method* used to perform the trick, and the *effect* that results from the method (for a detailed discussion on the stages involved in producing magical effects, see [55]). Crucially, the method refers to the *action*, or *combination of actions*, employed by magicians to lead spectators into a deceptive state, known as the *effect* (see e.g. [36]).

In magic tricks, spectators are deceived because they are disoriented by the magician's method. They expect the magician's action to result in some state of affairs but, as a result of the magician's action, they observe that another state of affairs obtain which contradicts their expectations. At the heart of the magician's deceitful effect is a mismatch between the spectators' expectations and the result they actually observe. Spectators expect the magician to do some particular action but in fact, the magician has not played this action but another one resulting in deception and surprise. Here, we use the French Drop Trick (see [5, 35]) to formalize and illustrate visual actions of misdirection. Accordingly, states of deception in the French Drop concern the method and, thus, the effect.

2.3.2 Casting the magic spell *logically*

We aim to define actions of misdirection of two distinct types. The first type is verbal and consists, for an operator, to induce a false belief in an addressee by making a linguistic utterance. The second type is visual and now consists to trigger a wrong observation (i.e., observing something that differs from reality) in the addressee by making gestures. Those types act at two different levels in addressees. In verbal misdirection, the operator's announcement acts at the level of the addressees' beliefs to make them believe the false, or at least to prevent them from believing the truth. In visual misdirection, the operator's gestures act at the level of the addressees' observations and beliefs, to make them see something that is not the case, as it happens in visual illusions for instance.

In this paper, the logic **DLM** will contain a static operator for expressing agents' beliefs as well as special atomic formulas for expressing their (non-epistemic) observations. But **DLM** will also contain a dynamic operator to capture the agents' informational actions. Semantically, we define specific action models for verbal actions, or actions of telling, and specific action models for visual actions, or actions of showing. In both cases, actions are either genuine or misleading.

Our setting turns out to be powerful enough to distinguish between simulation and dissimulation, in the verbal and visual cases. We argue that it can also express other notions such as epistemic observation and stronger notions of belief and observation. Our logic shows that visual actions convert atomic observation in epistemic observation but also that that simulation (of the false) implies dissimulation (of the truth). Finally, we use **DLM** as a tool to put more emphasis on the specificity of misdirection in magic tricks compared to other forms: the state of surprise caused by the deception involved. That way, we aim to capture the essence of misdirection in magic tricks as a novel application of Dynamic Epistemic Logic, but also to pave the way for using **DLM** to analyze other forms of misdirection in the future.

3 A Dynamic Logic of Misdirection

We present the syntax and semantics of our *Dynamic Logic of Misdirection* (**DLM**). We introduce a class of models called herein *observational epistemic models* that contain two kinds of information: about beliefs, which can result from both verbal communication and visual perception, and information about the facts that are observable for the agents, which result from visual perception only. While beliefs will be represented in the language via a standard doxastic modality [32], observations will be modelled by special atomic symbols, similarly to the way [4] models attention. Unlike other approaches like [31, 16], our framework will allow us to model dynamic aspects of beliefs and visual misdirection. We also discuss the appropriateness of certain classes of models and present a standard axiomatization for the logic.

3.1 Syntax and Semantics

The language presented herein is an adaptation of the Dynamic Epistemic Logic (DEL) for (in)attentive agents presented in [4], based on generalizing [2]. Throughout the text, we let Prop be a countable set of propositional symbols, Agt a finite set of agent symbols, and $\mathsf{Obs} = \{\mathsf{o}_a p, \mathsf{o}_a \neg p \mid a \in \mathsf{Agt},\ p \in \mathsf{Prop}\}$ a set of observation symbols. We read $\mathsf{o}_a p$ (respectively $\mathsf{o}_a \neg p$) as *"agent a observes that p (respectively $\neg p$) is the case"*. Notice that in [4], some similar extra symbols are used to state that an agent is paying attention to a certain atomic proposition p.

But here, the conceptual difference is that an extra atomic symbol $\mathsf{o}_a p$ is added to stand for non-epistemic observation, i.e., for the fact that an agent a only visually perceives that some fact p is happening, possibly without paying attention to it. We start out by formally introducing the models we use to interpret our logic.

Definition 1. A *relational (Kripke) model* is a tuple $\mathcal{M} = \langle W, \{R_a\}_{a \in \mathsf{Agt}}, V \rangle$ where:

- W is a non-empty set of *possible worlds*;
- $R_a \subseteq W \times W$;
- $V : \mathsf{Prop} \cup \mathsf{Obs} \to 2^W$ is a *valuation function.*

Let $w \in W$, we call the pair $(\mathcal{M}, w)$ a *pointed model*, with parentheses usually dropped.

Below we introduce a particular class of models that turns out to be interesting in the rest of the paper. Intuitively, in this class, we force the relations R_a to have the classical properties of *belief* [65, 32]. For observations, we need to impose conditions to guarantee that they are *consistent.* We will come back to this point later on the paper, after presenting the whole framework.

Definition 2. We define the class of *observational epistemic models* as the set of all the pointed models $\mathcal{M}, w$ such that $\mathcal{M} = \langle W, \{R_a\}_{a \in \mathsf{Agt}}, V \rangle$, $w \in W$, and where:

- each binary relation R_a is Euclidean, transitive and serial; and
- for all $p \in \mathsf{Prop}$ and $a \in \mathsf{Agt}$, $w \in V(\mathsf{o}_a p)$ implies $w \notin V(\mathsf{o}_a \neg p)$.

At this point, an important distinction to make concerns the models used for (in)attentive agents in [4], compared to the ones we use here. In [4], for each propositional symbol p, and each agent a, there is a special propositional symbol $\mathsf{h}_a p$ that labels a world in case the agent a is attentive to symbol p at such a world. In our setting, extra symbols for observation, written $\mathsf{o}_a p$ (with $p \in \mathsf{Prop}$), are no longer indicating "attentiveness", but they indicate that agent a is *observing* that a certain fact p holds, without necessarily paying attention to it. As already mentioned, the agent observes fact p in a non-epistemic way in this case (see e.g. [20, 21, 56, 19]), since the agent does not endorse any doxastic stance regarding the fact observed (e.g. when we see an entity in the water without forming any belief about it, for instance that it is a fish). In turn, we introduce two symbols instead of one: $\mathsf{o}_a p$ to indicate that the agent observes that p is the case, and $\mathsf{o}_a \neg p$ to indicate that the agent observes that p is not the case. As one can notice, $\mathsf{o}_a p$ being false is not the same as $\mathsf{o}_a \neg p$ being true since they can both be false but they cannot both be true at the same time. Finally, notice that each relation R_a is Euclidean, transitive and serial, i.e., has the usual (basic) properties of belief [65, 32].

The actions of misdirection that some agent(s) perform will be represented via standard tools of Dynamic Epistemic Logic known as *action models* [2]. Recall that action models are objects describing semantically the effects intended by an agent over an epistemic situation (i.e., over a model). They can be used as parameters in formulas of the language, in particular, inside a dynamic modality. In a nutshell, dynamic modalities perform updates in the current model, in this case depending on the semantic description given by an action model. When interpreting the dynamic modality, action models tell us how to update the original model, raising a new epistemic situation. In Definitions 3 and 4, action models and the language are defined as usual, by mutual recursion.

Definition 3. An *action model* is a tuple $\mathbf{A} = \langle E, \{\rightarrow_a\}_{a \in \mathsf{Agt}}, \mathsf{pre}, \mathsf{post}\rangle$, where:

- E is non-empty finite set of epistemic *actions* that, conceptually, can be either of verbal or visual type;
- $\rightarrow_a \subseteq E \times E$, for each $a \in \mathsf{Agt}$, is Euclidean, transitive and serial;
- $\mathsf{pre} : E \rightarrow \mathbf{DLM}$ is the *precondition function*, where **DLM** is the language given by Definition 4; and
- $\mathsf{post} : E \rightarrow (\mathsf{Prop} \cup \mathsf{Obs}) \rightarrow \mathbf{DLM}$, the *postcondition function*, is such that for a finite subset $D \subseteq \mathsf{Prop} \cup \mathsf{Obs}$, we have:
  - $\mathsf{post}(e)(p) \in \{\top, \bot\}$, for $p \in D$,
  - for all $\{\mathsf{o}_a p, \mathsf{o}_a \neg p\} \subset \mathsf{Obs}$, we have $\mathsf{o}_a p \in D$ if and only if $\mathsf{o}_a \neg p \in D$, and for all $e \in E$,
    * $\mathsf{post}(e)(\mathsf{o}_a p) = \top$ implies that $\mathsf{post}(e)(\mathsf{o}_a \neg p) = \bot$,
    * $\mathsf{post}(e)(\mathsf{o}_a \neg p) = \top$ implies that $\mathsf{post}(e)(\mathsf{o}_a p) = \bot$,
  - for all $p \notin D$, and all $e \in E$, $\mathsf{post}(e)(p) = p$.

Let $e \in E$, we call $(\mathbf{A}, e)$ a *pointed action model* (parentheses usually dropped).

Notice that in Definition 3, action models incorporate *postconditions*, as in [61], and unlike [4]. This variant does not add expressive power but simplifies the definition of actions of verbal and visual misdirection since the truth value of atomic observations can be updated. Postconditions are also required to update the models consistently, and respecting the properties of an observational model (see Lemma 16). In this regard, we only allow updates of the truth value of a finite set of symbols (either propositional or atomic observations) with $\top$ or $\bot$, and for the rest, post acts as the identity function. Moreover, we ask that if an observation $\mathsf{o}_a p$ is updated, so is its counterpart $\mathsf{o}_a \neg p$, and the update is defined accordingly. In the next definition, we make precise the syntax of our dynamic logic of misdirection.

Definition 4. The set of well-formed formulas of the language **DLM** is given by the following Backus-Naur Form:

$$\varphi, \psi ::= p \mid \mathsf{o}_a p \mid \mathsf{o}_a \neg p \mid \neg\varphi \mid \varphi \wedge \psi \mid \mathcal{B}_a \varphi \mid [\mathbf{A}, e]\varphi,$$

where $p \in \mathsf{Prop}$; $a \in \mathsf{Agt}$; and $\mathbf{A}, e$ is a pointed action model, whose preconditions and postconditions are defined in a lower level of the recursive definition. Other Boolean connectives are defined as usual: $\bot := p \wedge \neg p$, $\top := \neg\bot$, $\varphi \vee \psi := \neg(\neg\varphi \wedge \neg\psi)$, $\varphi \rightarrow \psi := \neg\varphi \vee \psi$ and $\varphi \leftrightarrow \psi := (\varphi \rightarrow \psi) \wedge (\psi \rightarrow \varphi)$. We also define the following abbreviations: $\hat{\mathcal{B}}_a \varphi := \neg\mathcal{B}_a \neg\varphi$; and $\langle \mathbf{A}, e \rangle \varphi := \neg[\mathbf{A}, e]\neg\varphi$. The set of literals is defined as $\mathsf{Lit} = \{\ell \mid \ell \in \mathsf{Prop} \text{ or } \ell = \neg p \text{ with } p \in \mathsf{Prop}\}$. If $\ell = \neg p$, we use sometimes $\neg\ell$ to denote p; and use $\mathsf{o}_a \neg\ell$ to denote $\mathsf{o}_a p$. Finally, we define the fragment **OL** to be the language generated from **DLM** when we remove formulas of the form $[\mathbf{A}, e]\varphi$.

The modalities introduced in the BNF above should be read as follows. Formula $\mathcal{B}_a \varphi$ means 'agent a believes that φ holds'. Its dual is $\hat{\mathcal{B}}_a \varphi$, meaning '$\varphi$ is believable for agent a', or, in other words, that 'φ is consistent with agent a's beliefs'. Here, by taking inspiration from [4], observations are not epistemic but factual. Thus, they are represented by special atomic propositions: $\mathsf{o}_a p$ indicates that *agent a observes that p is the case*, while $\mathsf{o}_a \neg p$ states that *agent a observes that p is not the case*. Definition 6 details the semantics of all the formulas of **DLM** but, intuitively, beliefs are conceived as weak doxastic attitudes, or so-called *prima facie* beliefs. Thus, when an agent a believes that φ, namely $\mathcal{B}_a \varphi$, agent a may lack any justification for believing φ. In Section 5, this notion of *weak belief* is used to define a notion of *strong belief* according to which agents hold some justification to support their beliefs. Finally, dynamic formulas $[\mathbf{A}, e]\varphi$ are read as 'after the action $\mathbf{A}, e$ is executed, φ holds', while its dual $\langle \mathbf{A}, e \rangle \varphi$ is read as 'the action $\mathbf{A}, e$ is executable at the current situation and after its execution φ will hold'.

Now, we define how an action model affects an observational epistemic model, — potentially modifying the doxastic states and observations of a set of agents. In general, these actions are acts of communication between agents which can consist in verbal announcements or visual gesture, and can be either genuine or deceptive. As in usual settings, such communication can be public, private or semi-private, but might involve both verbal and visual information herein. Moreover, notice that while in [4], actions of communication are *truthful*, we consider both truthful and untruthful actions here, as we want to address different forms of communication.

In addition, our action models are enriched with *postconditions*, which stand as descriptions of how certain facts change after an action, following e.g. [61]. In our case, postconditions are used to express changes in the facts that an agent is observing, as well as changes in the agent's observations themselves.

Definition 5. Let $\mathcal{M} = \langle W, \{R_a\}_{a \in \mathsf{Agt}}, V\rangle$ be a relational model, and let $\mathbf{A} = \langle E, \{\rightarrow_a\}_{a \in \mathsf{Agt}}, \mathsf{pre}, \mathsf{post}\rangle$ be an action model. We define their product $\mathcal{M} \otimes \mathbf{A} = \langle W', \{R'_a\}_{a \in \mathsf{Agt}}, V'\rangle$, where:

- $W' = \{(w, e) \in W \times E \mid \mathcal{M}, w \models \mathsf{pre}(e)\}$;
- $(w, e)R'_a(v, f)$ if and only if wR_av and $e \rightarrow_a f$, for all $a \in \mathsf{Agt}$;
- $V'(p) = \{(w, e) \mid \mathcal{M}, w \models \mathsf{post}(e)(p)\}$.

With all these definitions at hand, we can now define the semantics of **DLM**.

Definition 6. Let $\mathcal{M} = \langle W, \{R_a\}_{a \in \mathsf{Agt}}, V\rangle$ be a model, and let $w \in W$. The satisfaction relation $\models$ between $\mathcal{M}, w$ and formulas φ is defined as follows:

$$\begin{array}{lcl}
\mathcal{M}, w \models p & \Leftrightarrow & w \in V(p) \quad (p \in \mathsf{Prop} \cup \mathsf{Obs}) \\
\mathcal{M}, w \models \neg\varphi & \Leftrightarrow & \mathcal{M}, w \not\models \varphi \\
\mathcal{M}, w \models \varphi \wedge \psi & \Leftrightarrow & \mathcal{M}, w \models \varphi \text{ and } \mathcal{M}, w \models \psi \\
\mathcal{M}, w \models \mathcal{B}_a\varphi & \Leftrightarrow & \text{for all } v \in W \text{ such that } wR_av,\ \mathcal{M}, v \models \varphi \\
\mathcal{M}, w \models [\mathbf{A}, e]\varphi & \Leftrightarrow & \mathcal{M}, w \models \mathsf{pre}(e) \text{ implies } (\mathcal{M} \otimes \mathbf{A}), (w, e) \models \varphi.
\end{array}$$

A formula φ is satisfiable if there exists a pointed model $\mathcal{M}, w$ such that $\mathcal{M}, w \models \varphi$, and it is valid if $\mathcal{M}, w \models \varphi$, for all pointed models $\mathcal{M}, w$.

Remark 7. The dual modality $\langle \mathbf{A}, e\rangle$ of $[\mathbf{A}, e]$ is interpreted as:

$$\mathcal{M}, w \models \langle \mathbf{A}, e\rangle\varphi \Leftrightarrow \mathcal{M}, w \models \mathsf{pre}(e) \text{ and } (\mathcal{M} \otimes \mathbf{A}), (w, e) \models \varphi.$$

Let us make some observations concerning the execution of an action in our setting. Here we are interested in particular *"informational actions"*, in the sense that some agent a performs an informative act which leads to update the beliefs and observations of the rest of the agents. If agent a executes the action e, this action does not change her own factual beliefs and visual observation of the environment. However, other agents' beliefs and visual observations might change, but not necessarily though. This means that the agent can execute an action with some informational purpose but she cannot guarantee that this action will modify the other agents' beliefs and/or visual observations. Consider the following example: a magician performs an action to misdirect the attention of a crowd of spectators. The magician knows that actual facts do not change during the trick, but what she does

CPL	All axioms from propositional logic	
Obs	$\vdash \mathsf{o}_a p \to \neg \mathsf{o}_a \neg p$	$\dfrac{\vdash \varphi \quad \vdash \varphi \to \psi}{\vdash \psi}$ (MP)
$\mathtt{K}_{\mathcal{B}}$	$\vdash \mathcal{B}_a(\varphi \to \psi) \to (\mathcal{B}_a \varphi \to \mathcal{B}_a \psi)$	
$\mathtt{4}_{\mathcal{B}}$	$\vdash \mathcal{B}_a \varphi \to \mathcal{B}_a \mathcal{B}_a \varphi$	
$\mathtt{5}_{\mathcal{B}}$	$\vdash \neg \mathcal{B}_a \varphi \to \mathcal{B}_a \neg \mathcal{B}_a \varphi$	$\dfrac{\vdash \varphi}{\vdash \mathcal{B}_a \varphi}$ ($\mathtt{Nec}_{\mathcal{B}}$)
$\mathtt{D}_{\mathcal{B}}$	$\vdash \mathcal{B}_a \varphi \to \hat{\mathcal{B}}_a \varphi$	

Table 1: Axioms and rules for **OL**.

not know, and cannot guarantee, is whether the spectators will believe that actual facts *did* change. In other words, the magician cannot be sure that her misdirection will succeed. In formula, for an informational action model $(\mathbf{A}, e)$, we have that $\mathcal{M}, w \models \mathcal{B}_a \varphi$ does not imply $\mathcal{M}, w \models [\mathbf{A}, e]\mathcal{B}_a \varphi$.

Regarding the modelling of magic tricks, the approach of [57] using Propositional Dynamic Logic (PDL) differs from the DEL approach of **DLM** in terms of focus of interest. While both rely on modelling epistemic notions, such as beliefs and expectations, to express states of deceptions, the goal of **DLM** is to better understand the nature of misdirection. In particular, we are interested in visual misdirection, so the logic encodes notions such as atomic observation, as a non-epistemic attitude, and visual action of observation, now as an epistemic attitude, in addition to classical verbal announcements. By contrast, [57] puts more emphasis on the constructional aspects of conjuring magic.[1] They aim to express the various forms of *impossibility* spectators may experience with magical effects, based on characterizing the evidence relationship between expected events and real occurring events, –such as similarity, perceptual equivalence, structural equivalence, and congruence. To some extent, however, our two perspectives meet since, as we will see, **DLM** is used in subsection 5.4 to express the effect of *surprise* obtained with the French Drop trick.

3.2 Axiomatic System

We start out by introducing the axiom system OL for the static fragment **OL**.

Definition 8. The axiom system OL contains all the axioms and rules in Table 1.

Notice that in the presentation of the axiomatic system, we use axiom schemes instead of plain axioms, i.e., we use meta-variables for formulas instead of fixed propositional symbols (see e.g. [58] for details on the use of axiom schemes). Still, for simplicity's sake, we refer to axiom schemes just as *axioms* in what follows.

[1]Analyzing Martin Gardner's *Turnabout* trick (see [30]) in that respect.

As expected, axioms for $\mathcal{B}_a$ are classical for doxastic modalities. Axioms $4_{\mathcal{B}}$ and $5_{\mathcal{B}}$ make sense for beliefs obtained by verbal means or any kind of sensory activity other than observations: it makes sense that one believes what one actually believes but also believes that she disbelieves something when it is the case (i.e., they are the standard axioms of introspection). Axiom $\mathtt{D}_{\mathcal{B}}$ accounts for seriality, and means that what is believed by an agent must be consistent with the agent's beliefs. Concerning observation, we do not have axioms at the epistemic level, since our approach considers that agents may observe without necessarily believing. In our framework, an agent can observe that p, that is $\mathsf{o}_a p$, without p being true, as it happens to be the case in common illusions or hallucinations. However, as expected, an agent cannot observe p and $\neg p$ at the same time. What we need to establish is the inconsistency of observing a fact and its negation, given by axiom $\mathtt{Obs}$. Notice that this axiom is consistent with our basic requirement imposed on valuations in observational epistemic models (see Definition 2).

The proof of soundness and completeness follows standard arguments of modal logic (see, e.g., [17, 7, 8]). Soundness is straightforward, except for the axiom $\mathtt{Obs}$ which is novel. Notice that if a pointed observational epistemic model $\mathcal{M}, w$ is such that $\mathcal{M}, w \models \mathsf{o}_a p$, it is necessarily the case that $\mathcal{M}, w \not\models \mathsf{o}_a \neg p$, because $w \in V(\mathsf{o}_a p)$ implies $w \notin V(\mathsf{o}_a \neg p)$, by the second condition in Definition 2. Thus, the axiom $\mathsf{o}_a p \to \neg \mathsf{o}_a \neg p$ is also valid.

Following [7, Proposition 4.12], given an arbitrary consistent set of formulas Γ, completeness is obtained by building a model (known as the 'canonical model') satisfying all the formulas in Γ. Associated notions like derivability (via OL), theoremhood, maximal consistent sets (written MCS) and consistency/inconsistency are defined in the usual way.

Definition 9 (Canonical Model). The *canonical model* $\mathcal{M}^c = \langle W^c, \{R^c_a\}_{a \in \mathsf{Agt}}, V^c \rangle$ is defined as:

- $W^c := \{\Delta \mid \Delta \text{ is an MCS of } \mathsf{OL}\}$,
- $R^c_a := \{(\Delta_1, \Delta_2) \in W^c \times W^c \mid \text{ for all } \psi \text{ in } \mathbf{OL}, \mathcal{B}\psi \in \Delta_1 \text{ implies } \psi \in \Delta_2\}$, for each $a \in \mathsf{Agt}$, and
- $V(p) := \{\Delta \in W^c \mid p \in \Delta\}$, for each $p \in \mathsf{Prop}$.

The canonical model will be adequate as a model for any consistent set of formulas. But since this model is built over *maximal* consistent sets, the next classical lemma needs to be established.

Lemma 10 (Lindenbaum Lemma). *Let Γ be a consistent set. Then, there exists an MCS Γ' such that $\Gamma \subseteq \Gamma'$.*

Proof. See e.g. [7, Lemma 4.17]. □

Moreover, the canonical model defined above is adequate for our purpose, only if it is an observational epistemic model.

Lemma 11. *The canonical model* $\mathcal{M}^c = \langle W^c, \{R^c_a\}_{a\in\mathsf{Agt}}, V^c\rangle$ *from Definition 9 is an observational epistemic model.*

Proof. In order to prove this proposition, we have to show that each relation R^c_a is transitive, Euclidean and serial, and that for all $p \in \mathsf{Prop}$ and $a \in \mathsf{Agt}$, $\Delta \in V^c(\mathsf{o}_a p)$ implies $\Delta \notin V^c(\mathsf{o}_a \neg p)$. The properties of R^c_a are guaranteed by axioms $\mathtt{4}_{\mathcal{B}}$, $\mathtt{5}_{\mathcal{B}}$ and $\mathtt{D}_{\mathcal{B}}$ (see [7, Section 4.3]). Suppose that $\Delta \in V^c(\mathsf{o}_a p)$, for some MCS Δ, $a \in \mathsf{Agt}$ and $p \in \mathsf{Prop}$. By definition of V^c, $\mathsf{o}_a p \in \Delta$, and by $\mathtt{Obs}$, $\mathsf{o}_a p \rightarrow \neg\mathsf{o}_a \neg p \in \Delta$. Then, by $\mathtt{MP}$, $\neg\mathsf{o}_a\neg p \in \Delta$. By classical properties of MCS, $\mathsf{o}_a \neg p \notin \Delta$, which proves the property. □

Now we can link the notion of derivability in OL and the one of satisfiability (in the canonical model).

Lemma 12 (Truth Lemma)**.** *Let* $\mathcal{M}^c = \langle W^c, \{R^c_a\}_{a\in\mathsf{Agt}}, V^c\rangle$ *be the canonical model from Definition 9, and let* $\Delta \in W^c$*. Then, for all formulas* φ *of* **OL***:*

$$\varphi \in \Delta \text{ if and only if } \mathcal{M}^c, \Delta \models \varphi.$$

The proof of Lemma 12 proceeds by induction on the structure of φ. The case where φ is of the form $\mathcal{B}_a\psi$, is obtained using the inductive hypothesis and the definition of R^c_a. Once this property is established, completeness can be proved.

Theorem 13. *The axiom system* OL *is sound and strongly complete with respect to observational epistemic models.*

Proof. Suppose Γ is a consistent set of formulas. By Lemma 10, there exists a MCS Γ' such that $\Gamma \subseteq \Gamma'$. By Lemma 12, $\mathcal{M}^c, \Gamma' \models \Gamma$. Moreover, by Lemma 11, $\mathcal{M}^c$ is an observational epistemic model, which completes the proof. □

Now it is time to consider the full language **DLM**, i.e., to handle dynamic modalities as well. In order to axiomatize **DLM**, it is only needed to show that every formula $[\mathbf{A}, e]\varphi$ of **DLM** can be transformed into another formula φ' of **OL**, i.e., with no occurrences of the dynamic modality $[\mathbf{A}, e]$. This transformation is possible since dynamic modalities do not add expressive power to **OL** (see e.g. [64] on this aspect). That being done, we rely on the completeness of **OL** to establish the completeness of **DLM**.

We start out by introducing a list of valid formulas, usually called *reduction axioms*, that will be crucial to eliminate dynamic modalities. The next definition follows the presentation of [61], which also includes postcondition in action models.

Definition 14. The following equivalences are reduction axioms for eliminating the $[\mathbf{A}, e]$ modalities (by applying them on the inner occurrence of a dynamic modality):

$$\begin{array}{lll} \vdash [\mathbf{A}, e]p & \leftrightarrow & (\mathsf{pre}(e) \to \mathsf{post}(e)(p)) \\ \vdash [\mathbf{A}, e]\neg\varphi & \leftrightarrow & (\mathsf{pre}(e) \to \neg[\mathbf{A}, e]\varphi) \\ \vdash [\mathbf{A}, e](\varphi \wedge \psi) & \leftrightarrow & ([\mathbf{A}, e]\varphi \wedge [\mathbf{A}, e]\psi) \\ \vdash [\mathbf{A}, e]\mathcal{B}_a\varphi & \leftrightarrow & (\mathsf{pre}(e) \to \bigwedge\{\mathcal{B}_a[\mathbf{A}, f]\varphi \mid e \to_a f\}) \end{array}$$

Then, the next lemma (already established in [61, Section 2.4]) follows. The proof is similar to the proof for classical action models (see e.g., [64, Section 6.6]).

Lemma 15. *The reduction axioms from Definition 14 are valid.*

Before going further, let us discuss the properties of observational epistemic models in presence of an update via dynamic modalities. The next lemma states that the class of all observational epistemic models (i.e., those that are Euclidean, transitive and serial) is not closed under the execution of actions. However, if we give up seriality, the system is complete with respect to the obtained class.

Lemma 16. *Let $\mathcal{M}, w$ be a pointed epistemic observational model, and let $\mathbf{A}, e$ be an action model. Then,*

1. *$(\mathcal{M}, w)$ Euclidean, transitive and serial does not imply that $((\mathcal{M} \otimes \mathbf{A}), (w, e))$ is serial.*

2. *$(\mathcal{M}, w)$ Euclidean and transitive implies that $((\mathcal{M} \otimes \mathbf{A}), (w, e))$ is also Euclidean and transitive.*

Thus, as mentioned, an alternative to obtain completeness consists in relaxing the conditions imposed on the models, concerning the seriality condition in particular.

Definition 17. The axiom system DOL is defined as OL minus the axiom $\mathtt{D}_{\mathcal{B}}$, and extended with the reduction axioms from Definition 14, and the following rule for replacement of equivalents $\mathtt{RE}$:

$$\frac{\vdash \psi \leftrightarrow \psi'}{\vdash \varphi(p/\psi) \leftrightarrow \varphi(p/\psi')} \ (\mathsf{RE})$$

where $\varphi(p/\psi)$ is the result of replacing every occurrence of p in φ by ψ.

In order to eliminate all the occurrences of a dynamic modality in a formula (via reduction axioms), some form of replacement of equivalent formulas is needed. In this way, it is possible to apply the equivalences from Definition 14 as a rewriting system, pushing dynamic modalities inside of the formula and finally eliminating them. This is accomplished by the rule RE.

Theorem 18. *The system* DOL *is sound and strongly complete with respect to the class of relational models that are Euclidean and transitive.*

Proof. First, thanks to Lemma 15 (using the rule RE) we get that every formula of **DLM** is provable (in DOL) equivalent to one of **OL**. Thus, by Proposition 13 and Lemma 16, completeness over Euclidean and transitive relational models follows. □

Based on the technical issues described above (i.e., the loss of seriality after updating a model), we obtain some undesired conceptual behaviour. If some agent a believes that φ in a serial model (i.e., in which the current state has at least a successor via a), naturally it is not the case that the agent also believes $\neg\varphi$. This is due to the fact that some successor of the current point exists, and since all successors must satisfy φ, it cannot be the case that they also satisfy $\neg\varphi$. However, after updating a model via an action that does not preserve seriality (i.e., that removes all successors of the current state), the agent believes both φ and $\neg\varphi$, as not having successors trivializes what the agent believes. Therefore, the agent's beliefs may become *inconsistent*. Although this is an undesirable issue, it is a standard characteristic in dynamic epistemic logic, and thus we will not discuss this limitation any further here and choose to proceed with our framework despite it.

4 Misdirection via Action Types

We introduce a simple fragment of **DLM** that is enough to characterize the intended actions of misdirection, verbal and visual. Based on the notion of *action types* from [38], this fragment allows only four kinds of *action types* in dynamic modalities, that are either verbal or visual, and either truthful or untruthful in both cases.[2]

Remark 19. Both observational epistemic models and action models are labelled directed graphs, with nodes representing states, or worlds, in the former and atomic actions in the latter. To model the distinction graphically, we use circles to represent worlds of observational epistemic models, and squares to represent atomic actions of action models (as in Figures 2 and 3). We use double lines to point out the actual world, both in the observational epistemic models and in action models.

[2]More colloquial terms for truthful and untruthful *visual* actions would be "genuine" and "bogus" visual actions.

In observational epistemic models, when an atomic symbol p (from $\mathsf{Prop} \cup \mathsf{Obs}$) appears inside a circle representing a world w, it indicates that p holds at w. If the symbol p is not inside of a world v, it means that such a symbol does not hold at v.

In action models, while preconditions are explicitly displayed, postconditions are expressed as the conjuntion of the atomic symbols that are updated, meaning that the symbols become true if they appear in the affirmative form, and become false if they appear in the negated form. For the rest of the symbols that do not appear explicitly, the postcondition acts as the identity function.

4.1 Defining Action Types

Let us define the language $\mathbf{DLM}^{(+,-)}$ as the fragment of $\mathbf{DLM}$ where dynamic modalities are restricted to one of the following action types: **(1)** $\mathsf{tell}_a^+(\varphi)$, **(2)** $\mathsf{tell}_a^-(\varphi)$, **(3)** $\mathsf{show}_a^+(\psi)$, or **(4)** $\mathsf{show}_a^-(\psi)$, with φ an arbitrary $\mathbf{DLM}^{(+,-)}$-formula, and $\psi = \ell_1 \wedge \ldots \ell_n$, where each $\ell_i \in \mathsf{Lit}$, and $a \in \mathsf{Agt}$. Notice that for visual actions we only admit the action of showing a conjunction of literals. In what follows we say that $\ell \in \psi$, if $\ell = \ell_i$, for some $1 \leq i \leq n$. On the one hand, action type **(1)** indicates that agent a truthfully announces formula φ, whereas **(2)** indicates that a untruthfully announces that φ. On the other hand, in **(3)**, agent a executes a genuine visual action which aims addressees to observe every $\ell \in \psi$ when every ℓ actually holds, whereas in **(4)**, agent a executes a bogus visual action which aims addressees to observe that every $\ell \in \psi$ holds when none of them actually hold. As one can see, definitions of verbal and visual actions are not entirely symmetric since each has its own particularities. For instance, they differ regarding the kind of formulas the action applies to, with verbal announcements applying to arbitrary formulas and visual actions applying only to (conjunctions of) propositional literals. This is due to the fact that our framework focuses on propositional observation via the symbols in Obs. We introduce the formal definitions below.

Definition 20 (Verbal action types)**.** Let φ be a $\mathbf{DLM}$ formula, $a \in \mathsf{Agt}$ and let $B = \mathsf{Agt} \setminus \{a\}$. The action type $\mathsf{tell}_a^+(\varphi) = (\mathbf{A}, e)$, where $\mathbf{A} = \langle E, \{\rightarrow_a\}_{a \in \mathsf{Agt}}, \mathsf{pre}, \mathsf{post}\rangle$, is defined by:

- $E := \{e, f\}$,
- $\rightarrow_a := E \times E$,
- $\rightarrow_b := \{(e, f), (f, f)\}$, for all $b \in B$,
- $\mathsf{pre}(e) := \mathcal{B}_a\varphi$, and $\mathsf{pre}(f) := \varphi \wedge \mathcal{B}_a\varphi$, and
- $\mathsf{post}(g)(p) := p$, for all $g \in E$ and $p \in \mathsf{Prop} \cup \mathsf{Obs}$.

The action type $\mathsf{tell}^-_a(\varphi)$ is defined exactly as $\mathsf{tell}^+_a(\varphi)$, except that:

- $\mathsf{pre}(e) := \mathcal{B}_a \neg\varphi$.

Each of the action types in Definitions 20 and 21 is a pointed action model $(\mathbf{A}, e)$ such that some agent $a \in \mathsf{Agt}$ performs an informational action to the other agents (with e being some of its actions). Verbal actions update the accessibility relation of agent(s) b, more precisely, it restricts b's accessibilities to those states where formula φ holds, while maintaining agent's a accessibilities, and thus her beliefs, as they were. The action type $\mathsf{tell}^+(\varphi)$ corresponds to truthful announcement and is fairly standard, i.e., a believes that φ is the case and a's truthful announcement that φ restricts b's accessibilities to states where φ holds, such that b also believes that φ. Its untruthful counterpart $\mathsf{tell}^-(\varphi)$ (inspired by e.g. [62]) differs from $\mathsf{tell}^+(\varphi)$ in that b's accessibilities are still restricted to states where φ holds but now agent a (a.k.a., the performer) believes that $\neg\varphi$ holds, contrary to what she announces to b. Let us now consider the action types corresponding to visual events.

Definition 21 (Visual action types)**.** Let $\psi = \ell_1 \wedge \ldots \wedge \ell_n$ be a conjunction of literals, $a \in \mathsf{Agt}$ and let $B = \mathsf{Agt} \setminus \{a\}$. The action type $\mathsf{show}^+_a(\psi) = (\mathbf{A}, e)$, where $\mathbf{A} = \langle E, \{\to_a\}_{a \in \mathsf{Agt}}, \mathsf{pre}, \mathsf{post}\rangle$, is defined by:

- $E := \{e, f\}$,
- $\to_a := E \times E$,
- $\to_b := \{(e, f), (f, f)\}$, for all $b \in B$,
- $\mathsf{pre}(e) := \bigwedge_{\ell \in \psi}(\ell \wedge \mathsf{o}_a \ell)$, and $\mathsf{pre}(f) := \bigwedge_{\ell \in \psi} \neg \mathsf{o}_a \neg \ell$, and

for all $g \in E$,

- $\mathsf{post}(g)(p) := \top$, for all $p \in \{\ell, \mathsf{o}_b \ell \mid \ell \in \psi\}$,
- $\mathsf{post}(g)(q) := \bot$, for all $q \in \{\mathsf{o}_b \neg \ell \mid \ell \in \psi\}$, and
- $\mathsf{post}(g)(r) := r$, for all $r \in (\mathsf{Prop} \cup \mathsf{Obs}) \setminus \{\ell, \mathsf{o}_b \ell, \mathsf{o}_b \neg \ell \mid \ell \in \psi\}$.

The action type $\mathsf{show}^-_a(\psi)$ is defined exactly as $\mathsf{show}^+_a(\psi)$, except that:

- $\to_a := \{(e, e), (f, f), (f, e)\}$,
- $\to_b := \{(e, f), (f, f)\}$, for all $b \in B$.
- $\mathsf{pre}(e) := \bigwedge_{\neg\ell \in \psi}(\ell \wedge \mathsf{o}_a \neg \ell)$, and $\mathsf{pre}(f) := \bigwedge_{\ell \in \psi} \neg \mathsf{o}_b \neg \ell$, and

- $\mathsf{post}(e)(p) := \top$, for all $p \in \{\mathsf{o}_b\ell \mid \ell \in \psi\}$,
- $\mathsf{post}(e)(q) := \bot$, for all $q \in \{\ell, \mathsf{o}_b\neg\ell \mid \ell \in \psi\}$, and
- $\mathsf{post}(e)(r) := r$, for all $r \in (\mathsf{Prop} \cup \mathsf{Obs}) \setminus \{\ell, \mathsf{o}_b\ell, \mathsf{o}_b\neg\ell \mid \ell \in \psi\}$.
- $\mathsf{post}(f)(p) := \top$, for all $p \in \{\ell, \mathsf{o}_a\ell, \mathsf{o}_b\ell \mid \ell \in \psi\}$,
- $\mathsf{post}(f)(q) := \bot$, for all $q \in \{\mathsf{o}_b\neg\ell, \mathsf{o}_a\neg\ell \mid \ell \in \psi\}$, and
- $\mathsf{post}(f)(r) := r$, for all $r \in (\mathsf{Prop} \cup \mathsf{Obs}) \setminus \{\ell, \mathsf{o}_b\ell, \mathsf{o}_b\neg\ell, \mathsf{o}_a\ell, \mathsf{o}_a\neg\ell \mid \ell \in \psi\}$.

The first distinction between verbal and visual actions is the use of postconditions for visual actions (also known as 'effects', related to the public assignments of [60]). This is due to the fact that observations are handled atomically, so in order to update visual perception of agents, we have to update the truth values of the corresponding atomic symbols. The underlying action model in $\mathsf{show}_a^+(\psi)$ contains two actions. The "actual" action to be executed (named e) requires as a precondition that all the literals in ψ be true, and leads to update b's observations (for all $b \in B$) to observe the correct facts stated by ψ, and to not observe the opposite facts. The latter helps us to stay in the proper class of models after the update. The other possible action (f) only requires a to not observe the opposite of ψ, and updates b's observations as in e. It is worth noting that the accessibility relation in the announcement says that a may be uncertain about her own observations, while for b it is established that a does not observe the opposite to what she announces (which is enough for truthfully observing ψ).

The case of $\mathsf{show}_a^-(\psi)$ also deserves some explanation. The precondition of the actual action e now states that ψ is entirely false, and that agent a observes that ψ is false. Its postcondition updates b's observation, leading her to wrongly observe ψ as being true. In that respect, action f only requires b not to be able to observe the actual falsity of ψ, and to update her observations with ψ, making them wrong. Notice that those precondition and postcondition do not involve any doxastic notion of belief or uncertainty: accurate versus inaccurate notions of observation involved are strictly propositional, in the sense of Definition 4. Moreover, while in $\mathsf{show}_a^+(\psi)$ there is an edge labelled with a between e and f, this is not the case in $\mathsf{show}_a^-(\psi)$. The reason is that in the former, $\mathsf{pre}(f)$ involves a's observations (as ψ is a truthful announcement) while in the latter does not.

The action types just defined are displayed in Figures 2 and 3. In what follows, let $a \in \mathsf{Agt}$ and denote $B = \mathsf{Agt} \setminus \{a\}$. We consider ψ to be a conjunction of literals

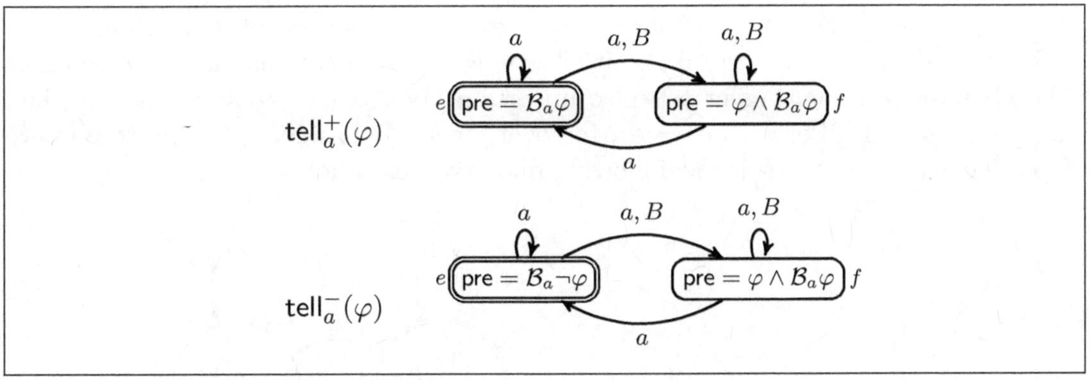

Figure 2: Types of Action Models Based on Distinct Modal Preconditions.

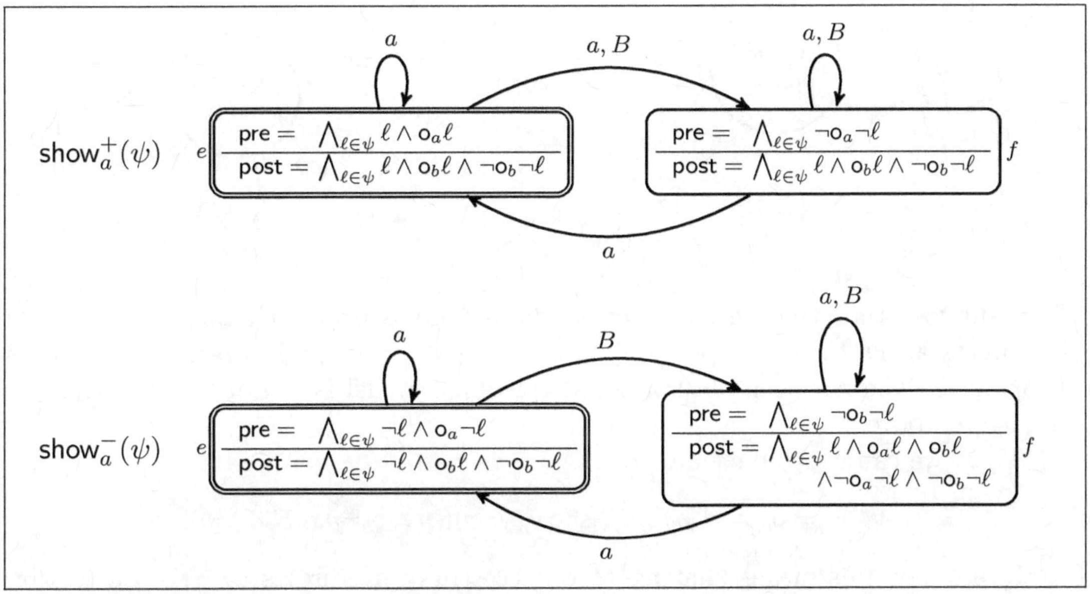

Figure 3: Types of Action Models Based on Distinct Observation-Based Preconditions.

and φ to be a **DLM**$^{(+,-)}$-formula. For simplicity's sake, we write only $o_b p$ and $o_b\neg p$ to indicate that the proposition p, respectively $\neg p$, holds for every $b \in B$. In verbal action types (see Figure 2), postconditions are the identity function, i.e., the action does not change the truth value of propositional symbols. In visual action types (see Figure 3), we update the value of propositional symbols involved in the visual action, as observation is handled at the level of literals (both from Prop and Obs). Example 22 illustrates the effects of the action types we defined.

Example 22. Consider the observational epistemic model displayed below on the left, in which agent a is certain about observing p, but agent b is uncertain whether she observes p or $\neg p$. Agent b is also uncertain about what agent a observes, i.e., whether agent a observes p or $\neg p$. Moreover, agent b considers possible to observe $\neg p$. The actual situation is the following one in which p holds:

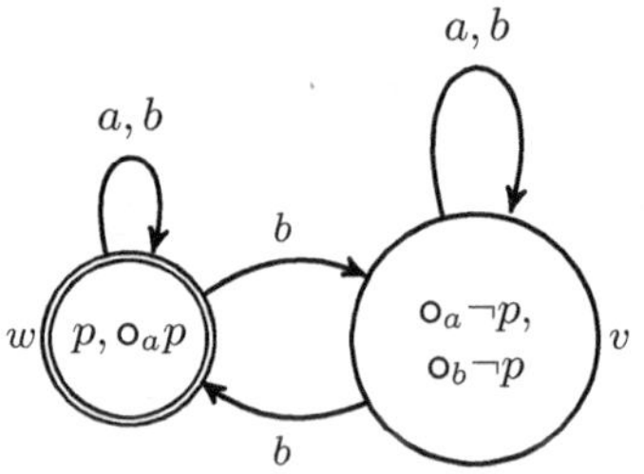

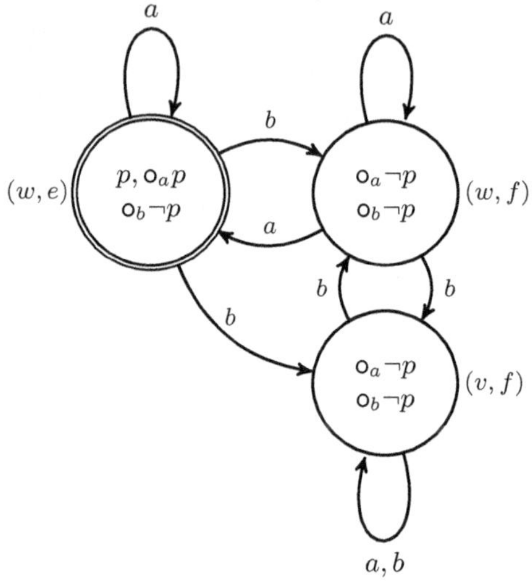

Now suppose that the dynamic action $\mathsf{show}^-_a(\neg p)$ is executed, i.e., that agent a misdirects agent b's visual perception, leading her to wrongly observe that $\neg p$ is happening. The resulting observational epistemic model is the one on the right in the figure above.

Thus, we can conclude that:

$$\mathcal{M}, w \models \mathsf{o}_a p \wedge \neg \mathsf{o}_b \neg p \wedge [\mathsf{show}^-_a(\neg p)](p \wedge \mathsf{o}_b \neg p \wedge \mathcal{B}_b \mathsf{o}_a \neg p).$$

Intuitively, this means that agent a successfully misdirects agent's b attention to $\neg p$: after a performs the action of showing $\neg p$, agent b (wrongly) observes $\neg p$ and believes that agent a also observes $\neg p$. Notice that we can also replace $[\mathsf{show}^-_a(\neg p)]$ by $\langle \mathsf{show}^-_a(\neg p) \rangle$, as the precondition of the action holds at the current state. If we want to force the action to be executed only in this case, the form $\langle \mathsf{show}^-_a(\neg p) \rangle$ should be used.

Remark 23. In this paper, we focus on what the spectators believe and observe about the magician's beliefs and observations. So far, we have not discussed the status of other second-order beliefs, such as what one spectator believes about another spectator's beliefs. Note that tell actions are, in fact, a particular form of semi-private announcements (see, e.g., [64, Chapter 6]); therefore, everything announced to a set of agents (here, the spectators) becomes common knowledge'[3]. Hence, after the tell actions, if $\mathcal{B}_b\mathcal{B}_a\varphi$ (with b a spectator and a the magician), then $\mathcal{B}_c\mathcal{B}_b\mathcal{B}_a\varphi$ (where c is also a spectator).

In Section 5, we will discuss how the previously defined action types enable us to characterize concepts such as simulation of the false and dissimulation of the truth. Moreover, we will discuss the logical implications of such formalizations.

4.2 Misdirection in the French Drop trick

With the machinery presented above, we can now formalize the French Drop trick as an illustration of the dynamic logic of misdirection **DLM**. Informally, the French Drop trick can be decomposed into three distinct steps:

1. First, there is the ***opening*** of the trick: the magician shows a coin to the public, shakes her hands and states that the coin is in one of her two hands, right or left.
2. Second, there is the ***method*** used by the magician, namely the actions she performs in front of the audience, and how those actions result in misdirection. On the one hand, there is an action of *simulation* through a fake passing of the coin. The magician seems to move the coin to her right hand but in fact the coin stays in her left hand. On the other hand, at the same time, there is an action of *dissimulation* through a genuine grabbing. The magician conceals the coin in her left hand by palming it. In the current section, we only model the action of simulation that takes place during the French Drop since, as we will show in subsection 5.3.2, dissimulation is a derived notion defined as the absence of simulation, verbal or visual. That being said, the magician can reinforce her effect by also making verbal announcements when passing and palming the coin, though not necessarily.
3. Finally, there is an ***effect*** produced by actions of simulation (*fake passing*) and dissimulation (*genuine palming*). The magician opens her right hand, and thus shows to the public that the coin is not in there, to the great surprise of the crowd.

Based on this informal description, we now model the French Drop trick in the fragment $\mathbf{DLM}^{(+,-)}$. The three steps of the magic trick (opening, method and

[3]We use common knowledge' because it is the standard terminology in Epistemic Logic, although, strictly speaking, we are concerned with beliefs here.

effect) are called the initial state, the epistemic action and the posterior state in the modelling herein.

- **Initial state:** we have two propositional symbols: l that stands for *"the coin is in the magician's left hand"*, and r for *"the coin is in the magician's right hand"*. Accordingly, we have four possible states, but the state in which the coin is at the same time in the magician's left and right hand is not actually plausible. Precisely, we have three options: either the coin is in the left hand $(l \wedge \neg r)$, the coin is in the right hand $(\neg l \wedge r)$, or there is no coin in the magician's hands $(\neg l \wedge \neg r)$. The actual state is the one in which the coin is in the magician's left hand: $(l \wedge \neg r)$. The following pointed observational epistemic model $\mathcal{M}, w$ models this opening situation:

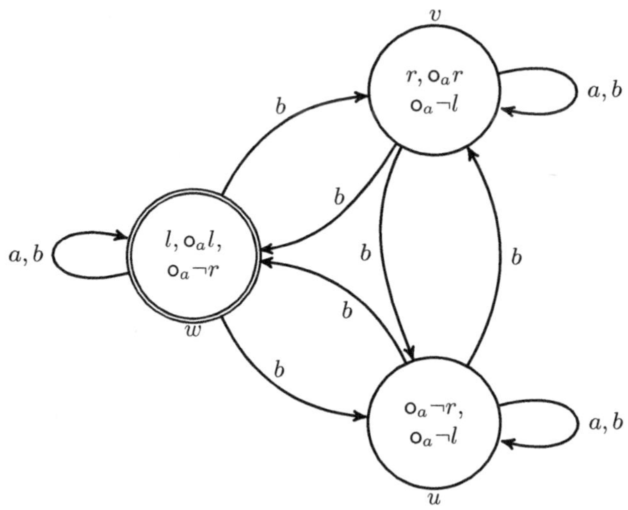

  Initially, only the magician believes that the coin is in her left hand. Moreover, on each possible situation, the magician, here represented by agent a, observes whether the coin is in a given hand and not in the other hand. For instance, in the actual state w, she observes that the coin is in her left hand, and that it is not in her right hand: $(l \wedge \neg r)$. But the audience is uncertain whether the coin is in the magician's left or right hand, and has not visual information. Let a be the agent modeling the magician, and w.l.o.g., let us assume there is only one agent b in the audience.

- **Visual actions:** we know that the magic actions are two-sided in the French Drop: on the one hand, it consists in pretending that r happens while, on the other hand, it consists in hiding that l remains the case. As indicated, we only model simulation (*fake passing*) here. This corresponds to action type $\mathsf{show}_a^-(r \wedge \neg l)$ and is illustrated below. With such an action, the magician aims at making the audience believe that the event r is taking place, when in fact, the magician herself believes and observes that it is not the case ($\mathcal{B}_a \neg r \wedge \mathsf{o}_a \neg r$). Also, it is obvious that the magician clearly observes and believes that the coin is still in her left hand ($\mathcal{B}_a l \wedge \mathsf{o}_a l$), as she is observing the real situation.

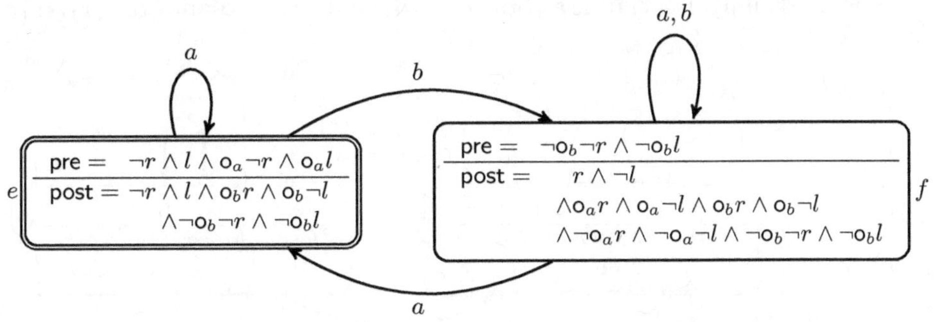

Below we depict the model obtained after executing the action $\mathsf{show}_a^-(r \wedge \neg l)$ in the opening scenario (here states (v, e) and (u, e) are ignored in the picture since $\mathcal{M}, v \not\models \mathsf{pre}(e)$ and $\mathcal{M}, u \not\models \mathsf{pre}(e)$):

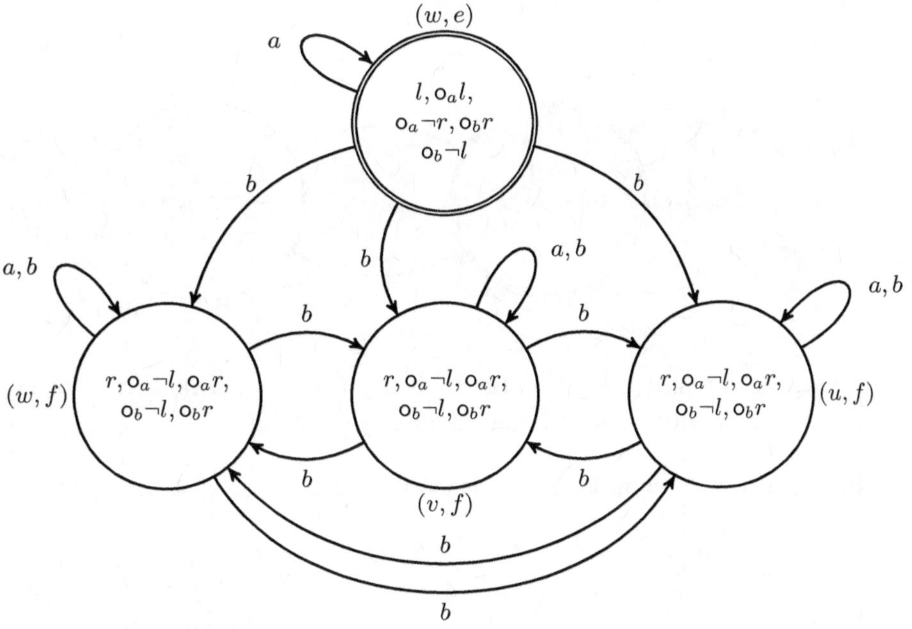

Notice that after this misdirection by simulation, all the states considered possible by agent b (the audience) become states in which she observes that the coin is in the magician's right hand. Observation became epistemic after the action: agent b also believes this (wrong) fact she is observing.

Finally, the execution of $\mathsf{show}_a^+(l \wedge \neg r)$ (depicted below) models the following fact: the magician reveals to the audience that they have been misdirected. This corresponds to the visual action of showing that the coin is actually in the magician's left hand, as opposed to what the audience observed previously.

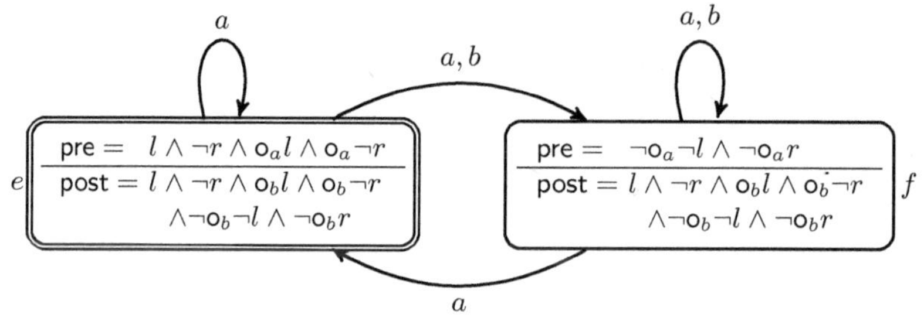

The result of executing the truthful visual action $\mathsf{show}_a^+(l \wedge \neg r)$ is depicted below:

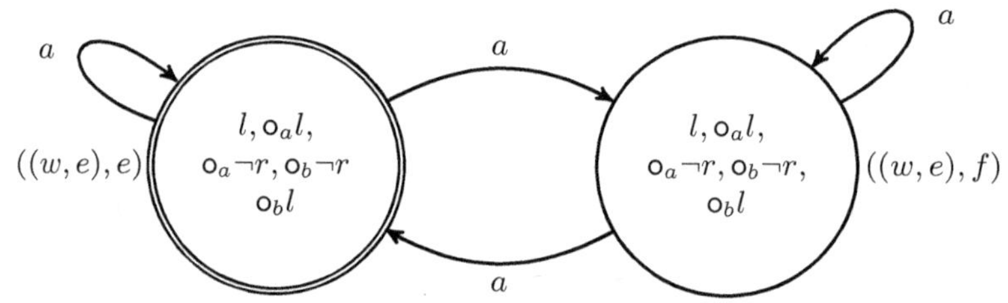

Again, as a result, agent b epistemically observes (this time, correctly) where the coin actually is, i.e. in the magician's left hand.

- **Posterior state:** naturally, l is still true. Formally, we get:

$$\begin{aligned} \mathcal{M}, w \models\ & l \wedge \mathsf{o}_a l \wedge \mathsf{o}_a \neg r \wedge \neg \mathsf{o}_a r \wedge \neg \mathsf{o}_b r \\ & \wedge \langle \mathsf{show}_a^-(r \wedge \neg l) \rangle (\mathsf{o}_a l \wedge \mathsf{o}_a \neg r \wedge \mathsf{o}_b r \wedge \mathsf{o}_b \neg l \\ & \qquad \wedge \langle \mathsf{show}_a^+(l \wedge \neg r) \rangle (\mathsf{o}_b l \wedge \mathsf{o}_b \neg r)). \end{aligned}$$

Each line in the formula above corresponds to one of the moments in which an action has been performed, and to their corresponding effects. In the first line, as described in the initial state, only the magician has a correct picture of the actual situation. Based on this initial state, as shown in the second line, the action of misdirection by simulation is executed, making b (wrongly) observe that the coin is in the magician's right hand ($\mathsf{o}_b r$) and no longer in her left hand ($\mathsf{o}_b \neg l$). Magician's beliefs and observations remain unchanged, as expected. When the trick is revealed, as shown in the third line, both the magician and spectators both have an adequate picture of the coin being in the magician's left hand (l) and not in the magician's right hand ($\neg r$). Their factual observations are correct: in the actual state (i.e. $((w, e), e)$), we have that $\mathsf{o}_a l$ and $\mathsf{o}_a \neg r$ hold for the magician; $\mathsf{o}_b l$ and $\mathsf{o}_b \neg r$ hold for the spectators. Their beliefs are correct too: $\mathcal{B}_a l$ and $\mathcal{B}_a \neg r$ for the magician; $\mathcal{B}_b l$ and $\mathcal{B}_b \neg r$ for the spectators. In addition to that, the magician and spectators have conscious beliefs of their own observations: it is the case that $\mathcal{B}_a(\mathsf{o}_a l)$ and $\mathcal{B}_a(\mathsf{o}_a \neg r)$, and also that $\mathcal{B}_b(\mathsf{o}_b l)$ and $\mathcal{B}_b(\mathsf{o}_b \neg r)$.

It is worth to notice that every action performed by the magician (in this case, $\mathsf{show}_a^-(r \wedge \neg l)$ and $\mathsf{show}_a^+(l \wedge \neg r)$) are executed only whenever their corresponding preconditions hold at their respective current situation, i.e., we use a diamond-like modality. This means that the magician *intentionally* performs these actions, in particular, she intentionally misdirects the spectators with $\mathsf{show}_a^-(r \wedge \neg l)$. By contrast, if we use a box-like modality (i.e., $[\mathsf{show}_a^-(r \wedge \neg l)]$), the formula will also hold when the precondition fails to be true; meaning that in case of misdirection, either the magician is not actually misdirecting the spectators, or that she unintentionally misdirects them. We focus herein on intentional actions, as discussed previously.

5 On the Generality of **DLM**

This section shows that **DLM** can be used to characterize more specific and general concepts, based on combining atomic observations, belief modalities and dynamic actions. We first define a stronger notion of belief, by reinforcing the basic beliefs of **DLM** with dynamic actions, and a stronger notion of observation, by using atomic observations to define epistemic observations. We then discuss how principles involving beliefs and observations interact, and why some of them should not be stated as axioms. Afterwards, we proceed to characterize the difference between misdirection as simulation (which involves an action) and misdirection as dissimulation (which involves the lack of an action). We finally attempt to capture the effect of surprise involved in the French Drop trick.

5.1 Stronger notions of Observation and Belief

The way logic **DLM** is interpreted in Definition 6, the notion of observation, written o, is atomic and non-epistemic. However, using a modal logic framework gives us the possibility to define a notion of epistemic observation, written $\mathcal{O}$, based on this initial concept of atomic (i.e., non-epistemic) observation o.

Definition 24. We define epistemic observation, written $\mathcal{O}$, as follows:

$$\mathcal{O}_b p \ := \ \mathsf{o}_b p \wedge \mathcal{B}_b(\mathsf{o}_b p).$$

According to this definition, agent a *epistemically observes* that p when agent b *atomically observes* that p, i.e. $\mathsf{o}_b p$, but also believes that she is atomically observing that p, i.e. $\mathcal{B}_b(\mathsf{o}_b p)$. That being done, we define a notion of epistemic observation which is *stronger* in the sense that it associates epistemic observations to dynamic actions.

Definition 25. We define strong epistemic observation, written $\mathcal{O}^{\mathsf{S}}$, as follows:

$$\mathcal{O}^{\mathsf{S}}_b p \ := \ \mathcal{O}_b p \wedge \bigvee_{a \in \mathsf{Agt} \setminus \{b\}} \langle \mathsf{tell}^+_a(p) \rangle \mathcal{B}_b p \vee \langle \mathsf{show}^+_a(p) \rangle \mathcal{B}_b p$$

Strong epistemic observation should be understood as epistemic observations that turn out to be supported by verbal and/or visual actions at the same time.

In contrast with our initial notion of belief $\mathcal{B}$, another notion we can define is *strong belief*. As it is, the modality $\mathcal{B}$ defines belief as being weak doxastic attitude in which the agent may lack any justification for supporting her beliefs. On the contrary, we define strong beliefs as being weak beliefs supported by genuine actions, either verbal or visual, as in the following definition.

Definition 26. Let φ be a conjunction of literals, we define strong beliefs, written $\mathcal{B}^{\mathsf{S}}$, in the following way:

$$\mathcal{B}^{\mathsf{S}}_b \varphi \ := \ \mathcal{B}_b \varphi \wedge \bigvee_{a \in \mathsf{Agt} \setminus \{b\}} \langle \mathsf{tell}^+_a(\varphi) \rangle \mathcal{B}_b \varphi \vee \langle \mathsf{show}^+_a(\varphi) \rangle \mathcal{B}_b \varphi$$

This definition means that agent b believes that φ in a strong sense when b 'weakly' believes that φ and this belief is robust under any truthful action with φ, verbal ($\langle \mathsf{tell}^+_a(\varphi) \rangle$) or visual ($\langle \mathsf{show}^+_a(\varphi) \rangle$). In other words, agent b holds a strong belief that φ because b's belief is supported by the possibility of agent a's action.

5.2 Interaction between Beliefs and Observations

Let us discuss some principles linking observations and beliefs, that are either satisfiable, valid or unreasonable in our setting. The first principle, hereafter separated into (1a) and (1b), states that if an agent a observes a certain fact as true (respectively, as false), a believes her observations to be true (respectively, false):

(1a) $\mathsf{o}_a p \rightarrow \mathcal{B}_a \mathsf{o}_a p$

(1b) $\mathsf{o}_a \neg p \rightarrow \mathcal{B}_a \mathsf{o}_a \neg p$

Principles (1a)-(1b) are *satisfiable* but not valid in our setting. As indicated before, we aim to keep a distinction between beliefs and observations since observation consists in sensory visual perception that does not necessarily involve conscious representation (for instance, you may see an entity in the water without seeing it *as* a fish in the water). So, if we accept principles (1a) and (1b) as universally valid, observations and belief conflate into the concept of "epistemic observation" discussed in Definition 24. Naturally, these statements are satisfiable: it can be the case that an agent observes a fact p (respectively, $\neg p$) and that she believes she is observing p (respectively, $\neg p$).

The following principle, now divided into (2a) and (2b), states that an agent cannot have wrong beliefs about its observations:

(2a) $\mathcal{B}_a \mathsf{o}_a p \rightarrow \mathsf{o}_a p$

(2b) $\mathcal{B}_a \mathsf{o}_a \neg p \rightarrow \mathsf{o}_a \neg p$

Principles (2a)-(2b) are *satisfiable* since there are states in which an agent has correct beliefs about her observations. That being said, they are not valid in our setting since, otherwise, an agent would never be misdirected, contrary to our purposes. To conclude, consider the following principle, separated into (3a) and (3b):

(3a) $(\mathsf{o}_a p \wedge \mathcal{B}_a \mathsf{o}_a p) \rightarrow [\mathsf{show}_a^+(p)](\mathsf{o}_b p \wedge \mathcal{B}_b \mathsf{o}_b p)$

(3b) $(\mathsf{o}_a \neg p \wedge \mathcal{B}_a \mathsf{o}_a \neg p) \rightarrow [\mathsf{show}_a^-(p)](\mathsf{o}_b p \wedge \mathcal{B}_b \mathsf{o}_b p)$

Principles (3a)-(3b) are now *valid* in our framework. Instead of relating beliefs and observations statically, they make use of the dynamic actions in order to guarantee that an observation is believable. Formula (3a) states that whenever agent a epistemically observes that p, then, if a truthfully shows that p to some addressee b, then b will also come to epistemically observe that p too. Formula (3b) states the same success but with simulation that p: whenever agent a epistemically observes that $\neg p$, then, if a now untruthfully shows that p to some addressee b, then b will come to epistemically observe that p. These principles reflect properly the spirit of the **DLM** framework.

5.3 Other Variations on Misdirection

5.3.1 Misdirection as Simulation

Having defined agent a's verbal simulation that $\neg p$ with $\mathsf{tell}_a^-(\neg p)$ and agent a's visual simulation that $\neg p$ with $\mathsf{show}_a^-(\neg p)$, we can define agent a's general simulation that $\neg p$ by the disjunction of both verbal and visual types, as follows.

Definition 27. We define simulation of the false, written Sim for simulation, by the following disjunction:

$$\mathsf{Sim}_a\neg p \ := \ (\mathcal{B}_a p \wedge \langle \mathsf{tell}_a^-(\neg p)\rangle \mathcal{B}_b \neg p) \vee (\mathsf{o}_a p \wedge \langle \mathsf{show}_a^-(\neg p)\rangle \mathsf{o}_b \neg p)$$

where $p \in \mathsf{Prop}$, and $\mathsf{Sim}_a\neg p$ expresses that *"agent a simulates that $\neg p$ is the case to agent b"*.

Definition 27 expresses simulation of a factual information $\neg p$ as being either verbal simulation of $\neg p$ (left-hand disjunct with tell^-) or visual simulation of $\neg p$ (right-hand disjunct with show^-). Since the formula $\mathsf{Sim}_a\neg p$ is a disjunction of cases, it is sufficient that either verbal simulation obtain or that visual simulation obtain for $\mathsf{Sim}_a\neg p$ itself to obtain. In the left-hand disjunct of $\mathsf{Sim}_a\neg p$, verbal simulation is expressed by the conjunction $\mathcal{B}_a p \wedge \langle \mathsf{tell}_a^-(\neg p)\rangle \mathcal{B}_b \neg p$. This means that agent a verbally simulates that $\neg p$ to agent b when a believes that p but a tells b that $\neg p$ such that, as a result of the verbal action $\mathsf{tell}_a^-(\neg p)$, agent b believes that $\neg p$. In the right-hand disjunct of $\mathsf{Sim}_a\neg p$, now, visual simulation is defined by the conjunction $\mathsf{o}_a p \wedge \langle \mathsf{show}_a^-(\neg p)\rangle \mathsf{o}_b \neg p$. This means that agent a visually simulates that $\neg p$ to agent b when a visually observes that p and, as a result of the visual action $\mathsf{show}_a^-(\neg p)$, agent b observes that $\neg p$. In both verbal and visual simulation of $\neg p$, the aim of a is that b holds a wrong visual representation that $\neg p$ through the action performed. Moreover, we use again diamond-like dynamic modalities since we aim to model the actions performed by the magician are intentional.

5.3.2 Misdirection as Dissimulation

So far, we have used **DLM** to express the notion of simulation, or faking, in the verbal case (tell^-) and in the visual case (show^-). The operator, e.g. the magician, uses a verbal or visual action to direct the addressees' attention, e.g. the spectators', toward a false representation of reality. As noticed in Section 2, however, this kind of misdirection based on an effective action differs from any other kind of misdirection, known as *"dissimulation"*, or hiding, in which no action takes place in order to deceive. In this case, deception *does not* result from any action, but from the failure of a verbal or visual action. Quite fortunately, **DLM** is expressive enough to also capture such a form of information dissimulation consisting in withholding information.

Definition 28. We define dissimulation of information, written Dis, by the following disjunction:

$$\mathsf{Dis}_a p := (\mathcal{B}_a p \land \mathcal{B}_b \neg p \land \neg\langle \mathsf{tell}_a^+(p)\rangle \mathcal{B}_b \neg p) \lor (\mathsf{o}_a p \land \mathsf{o}_b \neg p \land \neg\langle \mathsf{show}_a^+(p)\rangle \mathsf{o}_b \neg p)$$

where $p \in \mathsf{Prop}$, and $\mathsf{Dis}_a p$ expresses that *"agent a dissimulates that p is the case to agent b"*.

Definition 28 expresses dissimulation of a factual information p as being either verbal dissimulation of p (left-hand disjunct referring to tell^+) or visual dissimulation of p (right-hand disjunct referring to show^+). Since the formula $\mathsf{Dis}_a p$ is a disjunction of cases, it is sufficient that either verbal dissimulation obtain or that visual dissimulation obtain for $\mathsf{Dis}_a p$ itself to obtain. In the left-hand disjunct of $\mathsf{Dis}_a p$, verbal dissimulation is expressed by the conjunction $\mathcal{B}_a p \land \mathcal{B}_b \neg p \land \neg\langle \mathsf{tell}_a^+(p)\rangle \mathcal{B}_b \neg p$. Here, this means that agent a verbally hides that p to agent b when a believes that p and b believes that $\neg p$. But crucially, contrary to Definition 27 above, now *no verbal action* $\mathsf{tell}_a^+(p)$ has participated to make agent b actually believe that $\neg p$, i.e. not believe that p. In other words, agent b believes that $\neg p$ *by the lack of an action.* In the right-hand disjunct of $\mathsf{Dis}_a p$, now, visual dissimulation is defined by the conjunction $\mathsf{o}_a p \land \mathsf{o}_b \neg p \land \neg\langle \mathsf{show}_a^+(p)\rangle \mathsf{o}_b \neg p$. This means that agent a visually hides that p to agent b when a visually observes that p, b observes that $\neg p$ and here again, crucially, *no visual action* $\mathsf{show}_a^+(p)$ has participated to make agent b observe $\neg p$, i.e., not observe p. In both verbal and visual dissimulation of p, the aim of a is that b hold the wrong representation that $\neg p$, not as the result of any action but, specifically, by the lack of an action.

Recall that in the French Drop trick, dissimulation is not omission of information, or withholding information, but hiding information. The magician performs some visual action, i.e. some hand gesture, to make the audience wrongly observe that the coin moved to her right hand. In our own variant of the trick, however, the magician finally reveals her secret at the end of the trick: she opens her left hand in front of the public. But if the magician had not revealed her secret, this would have counted as a case of dissimulation in the sense of withholding, or omitting, information. In doing so, more precisely in doing nothing to correct the spectators' mistaken observations, the magician would have omitted visual information from them. The strength of logic **DLM** is to be able to tease apart dissimulation via hiding from dissimulation via omission.

5.3.3 Simulation with Dissimulation

Simulation of the false is a notion conceptually distinct from dissimulation of the truth (see subsection 2.2). We have used **DLM** to make this difference explicit by distinguishing simulation (Sim) from dissimulation (Dis). But though simulation and dissimulation are conceptually distinct, they are also logically related.

Intuitively, an agent who makes it appear that $\neg p$ (simulation) thereby also conceals the true state of affairs that p (dissimulation). Accordingly, in logic **DLM**, if agent a simulates the falsity of p, either verbally by performing the action $\mathsf{tell}_a^-(\neg p)$ or visually by performing the action $\mathsf{show}_a^-(\neg p)$, agent a also comes to dissimulate the truth of p at the same time. This is the case in the French Drop trick when the magician performs the visual action of simulating that $(\neg r \wedge l)$ (i.e. that the coin is not in the magician's right hand but in the magician's left hand): $\mathsf{show}_a^-(r \wedge \neg l)$. Naturally, at the same time, the magician comes to not show, i.e. to dissimulate, the truth of $(\neg r \wedge l)$: no execution of $\mathsf{show}_a^+(\neg r \wedge l)$ reveals such value (see subsection 4.2).

Proposition 29. *Let $p \in \mathsf{Prop}$ and let $\mathcal{M}, w$ be a pointed model, then:*

1. ***Verbal case:*** $\mathcal{M}, w \models \langle \mathsf{tell}_a^-(\neg p)\rangle \mathcal{B}_b \neg p \rightarrow \neg \langle \mathsf{tell}_a^+(p)\rangle (\hat{\mathcal{B}}_b \top \wedge \mathcal{B}_b \neg p)$;

2. ***Visual case:*** $\mathcal{M}, w \models \langle \mathsf{show}_a^-(\neg p)\rangle o_b \neg p \rightarrow \neg \langle \mathsf{show}_a^+(p)\rangle o_b \neg p$.

Proof. 1) For the verbal case, let us suppose that $\mathcal{M}, w \models \langle \mathsf{tell}_a^-(\neg p)\rangle \mathcal{B}_b \neg p$. By the definition of $\models$, we have that: (a) $\mathcal{M}, w \models \mathcal{B}_a p$, and that (b) $\mathcal{M}', (w, e) \models \mathcal{B}_b \neg p$, where $\mathcal{M}' = \mathcal{M} \otimes \mathsf{tell}_a^-(\neg p)$. Then, for agent b (the spectator) the only possible world after the update, is one in which the action f from $\mathsf{tell}_a^-(\neg p)$ was executed, i.e., a world that initially satisfied formula $(\neg p \wedge \mathcal{B}_a \neg p)$. Thus, for every possible successor of (w, e) via agent b, formula $\neg p$ holds (notice that since p is atomic, we can be sure there is no Moorean phenomena there[4]). To verify that $\mathcal{M}, w \models \neg \langle \mathsf{tell}_a^+(p)\rangle (\hat{\mathcal{B}}_b \top \wedge \mathcal{B}_b \neg p)$, we aim for a contradiction, thus we assume $\mathcal{M}, w \models \langle \mathsf{tell}_a^+(p)\rangle (\hat{\mathcal{B}}_b \top \wedge \mathcal{B}_b \neg p)$. By $\models$, we have that: (a') $\mathcal{M}, w \models \mathcal{B}_a p$, and that (b') $\mathcal{M}'', (w, e) \models (\hat{\mathcal{B}}_b \top \wedge \mathcal{B}_b \neg p)$, where $\mathcal{M}'' = \mathcal{M} \otimes \mathsf{tell}_a^+(p)$. Naturally, (a') holds as a consequence of (a), thus guaranteeing the executability of the actions.

Let us analyze (b'). If (w, e) has no successors via agent b in $\mathcal{M}''$, it would contradict $\hat{\mathcal{B}}_b \top$. Then, since we force (w, e) to have a successor via agent b, in such a case, $\mathcal{B}_b \neg p$ being true at (w, e) would imply that each of its successors satisfy $\neg p$. But again, the only possible successor via agent b would be one in which f satisfies p (by the definition of $\mathsf{tell}_a^+(p)$), thus we have a contradiction. Therefore, $\mathcal{M}, w \models \langle \mathsf{tell}_a^-(\neg p)\rangle \mathcal{B}_b \neg p \rightarrow \neg \langle \mathsf{tell}_a^+(p)\rangle (\hat{\mathcal{B}}_b \top \wedge \mathcal{B}_b \neg p)$.

[4]For a more extensive discussion about the Moorean phenomena, see e.g. [33].

2) The reasoning is similar for the visual case, except that notice that $\mathsf{o}_b\neg p$ holds after executing $\mathsf{show}^-(\neg p)$ in the current point of evaluation and does not hold after executing $\mathsf{show}^+(p)$, because of the postconditions of such action types. □

Notice that in the validities above, all informational actions are done at a propositional level (i.e., not complex formulas). This can be seen as a limitation of the established property, but in fact it can be generalized to conjunctions of literals. Actually, we use this approach to model the French Drop trick in subsection 4.2.

Then, we obtain a result we expect, based on Definitions 27 and 28:

Corollary 30. *For all $p \in \mathsf{Prop}$, and for all $\mathcal{M}, w$, the following statement holds:*

$$\mathcal{M}, w \models (\mathsf{Sim}_a\neg p \wedge (\mathcal{B}_b\neg p \wedge \mathsf{o}_b\neg p)) \rightarrow \mathsf{Dis}_a p.$$

Corollary 30 states that simulation of $\neg p$ implies dissimulation of p, but only under the assumption that agent b both believes and observes $\neg p$ (this ensures that simulation of $\neg p$ to agent b actually succeeds since b also believes and observes $\neg p$). More precisely, it states that when agent a simulates that $\neg p$ *successfully* in the sense that agent b believes and observes that $\neg p$, this implies that agent a also comes to dissimulate that p at the same time. This implication can be understood as follows in the French Drop trick: if the magician succeeds in making the spectators wrongly believe and observe that the coin has moved from her left to right hand, then the magician has succeeded in dissimulating that the coin has remained in her left hand all along. In other words, simulating successfully that the coin has moved from left to right hand implies to dissimulate that the coin in fact remained in the magician left hand. Notice that this holds only in the case that the spectators believed and observed $\neg p$ before, otherwise dissimulation naturally may not succeed.

The situation just described comes from the fact that we consider that the magician always succeeds in the French Drop trick (under the appropriate circumstances), so the logical behaviour is actually consistent with this reasonable assumption. Weaker actions and connections between simulation and dissimulation are worth to be explored in the future, e.g., if we consider a clumsy magician that fails in some of her actions.

5.4 Magic Misdirection as a Source of Surprise

We have seen that logic **DLM** is strong enough to express various notions characterizing misdirection in general and in magic tricks, more specifically. But in fact, **DLM** is also strong enough to express the sense of surprise raised by magic effects on spectators.

In the French Drop trick, it holds that $\langle \mathsf{show}^-_a(r \wedge \neg l)\rangle(\mathsf{o}_a l \wedge \mathsf{o}_a \neg r \wedge \mathsf{o}_b \neg l \wedge \mathsf{o}_b r)$, at an intermediary step. So, after a visual action of simulation $\mathsf{show}^-_a(r \wedge \neg l)$, spectators observe that $\neg l$ and r: $\mathsf{o}_b \neg l \wedge \mathsf{o}_b r$. Immediately after this effect, the magician unveils the mystery of the trick with visual action, i.e., it holds that $\langle \mathsf{show}^+_a(l \wedge \neg r)\rangle(\mathsf{o}_b l \wedge \mathsf{o}_b \neg r)$. In words, after the execution of $\mathsf{show}^+_a(l \wedge \neg r)$, the spectators observe the true facts l and $\neg r$: $\mathsf{o}_b l \wedge \mathsf{o}_b \neg r$. As one can see, those facts are completely opposite to the facts they observed just before.

Following [45, 44, 39], an agent is surprised by an incoming event when she recognizes an inconsistency, that is to say a *discrepancy* or *mismatch*, between her expectations about the world and the actual state she observes. Quite intuitively, the wider the gap between the agent's expectations and reality, the stronger the surprise. In that respect, [40] distinguishes a first-order kind of surprise called "*mismatch-based surprise*" from a second-order kind of surprise called "*astonishment*", or "*surprise in recognition*".

Mismatch-based surprise results from a conflict between the agent's scrutinized representation and perceived facts or events. The agent is surprised because she has some anticipatory representation, but she cannot make incoming events fit with this representation. The strength of the surprise effect depends on the agents' expectations, more precisely *beliefs* about future events. Surprise occurs if an unforeseen event, in the sense of *not believed* to happen, ultimately occurs. But a stronger surprise occurs if an event that was *believed not* to happen ultimately happens. In the **DLM** framework, for a given fact p, this can be expressed by the difference between agent a not believing that p when p happens, i.e. $\neg \mathcal{B}_a p \wedge p$, versus agent a believing that $\neg p$ when p happens: $\mathcal{B}_a \neg p \wedge p$. Since the agent's disbelief that p is stronger in the second case, surprise will also be stronger.

Astonishment, or surprise in recognition, results from the *recognition* of the absolute implausibility of a perceived fact compared to expectations. In this case, surprise is rooted in incongruity: an event not conceived as a relevant possibility actually takes place and leaves agents completely astonished. In **DLM**, agent a will be astonished that a fact p is true when p is true but agent a actually neither believes it to be true nor believes it to be false: $p \wedge \neg \mathcal{B}_a p \wedge \neg \mathcal{B}_a \neg p$. In other words, astonishment results from the fact that when p occurs, p is an epistemic blindspot.

In the French Drop trick, the surprise the public experiences is of the mismatch-based kind. When action $\mathsf{show}^-_a(r \wedge \neg l)$ leads spectators to atomic (i.e., non-epistemic) observations $\mathsf{o}_b \neg l$ and $\mathsf{o}_b r$, it also leads them to epistemically observe the same facts. The reason is that $\mathsf{show}^-_a(r \wedge \neg l)$ results in $\mathcal{B}_a(\mathsf{o}_b \neg l)$ and $\mathcal{B}_a(\mathsf{o}_b r)$, so by definition of epistemic observation (see subsection 5.1), we have $\mathcal{O}_b(\neg l)$ and $\mathcal{O}_b(r)$. At the end of the trick, the magician dispels her secret with visual action $\mathsf{show}^+_a(l \wedge \neg r)$ now leading the spectators to atomically observe l and $\neg r$: $\mathsf{o}_b l$ and

$\mathsf{o}_b\neg r$. Again, spectators come to epistemically observe the same facts since action $\mathsf{show}_a^+(l \wedge \neg r)$ results in $\mathcal{B}_a(\mathsf{o}_b l)$ and $\mathcal{B}_a(\mathsf{o}_b\neg r)$, so by definition of epistemic observation, we have: $\mathcal{O}_b(l)$ and $\mathcal{O}_b(\neg r)$. At the end of the trick, the spectators' epistemic observations $\mathcal{O}_b(l)$ and $\mathcal{O}_b(\neg r)$ are completely opposite to their epistemic observations $\mathcal{O}_b(\neg l)$ and $\mathcal{O}_b(r)$ one moment earlier, resulting in mismatch-based surprise.

Notice that, according to [40], astonishment would have replaced mismatch-based surprise if, when revealing the trick to the audience, the magician had shown, for instance, a banknote instead of a coin, like a rabbit straight out of her hat.

6 Conclusion

This article aims to conduct a logical analysis of misdirection, guided by the conceptual distinction between *non-epistemic* and *epistemic seeing* [20, 21], as well as the classical contrast between *simulation* and *dissimulation* [6]. We adapt extant work in dynamic epistemic logic [2, 64] and in logics of attention and observation [4, 11] to present a dynamic logic of misdirection, denoted **DLM**. In this way we provide a novel application of (a variant of) Dynamic Epistemic Logic in the formal analysis of both verbal and visual misdirection. The static part of the logic relies on atomic formulas for representing the agents' visual observation of their environments, and on classical doxastic modality for representing their beliefs. For the dynamic part, we focus on a fragment named $\mathbf{DLM}^{(+,-)}$, in which action types tell and show are defined for expressing verbal actions of announcing information, as well as visual actions of showing information. Those actions can be either truthful (indicated by label $^+$) or untruthful (indicated by label $^-$), making it possible to represent actions of verbal and visual misdirection. We used **DLM** to model a magic trick called the "French Drop" which illustrates actions of visual simulation (*fake passing*) and of visual dissimulation (*genuine palming*). The framework proved to be powerful enough to express the various notions involved in misdirection but also more complex notions, such as stronger belief attitudes, epistemic observations and the logical interactions between simulation and dissimulation. Interestingly, the setting also succeeded at capturing the specific sense of surprise resulting from magical effects.

We argue that our framework sets the basis for a broader formal theory of misdirection. Epistemic and modal logics in general are suitable to model other forms of misdirection, such as withholding information or conspiration between a set of agents. For instance, we can represent collective acts of misdirection (i.e., a magician having accomplices) by generalizing informational actions with respect to more than one agent, by using e.g. *private announcements*. On the other hand, it would be interesting to investigate different conditions on the modality of observation. In this work, we only ask for the underlying relation to be serial, in correspondence with the D axiom, while other constraints might be imposed (see e.g. [12, 13] concerning non-transitive knowledge used for modeling perceptual indistinguishability). Moreover, the classes of models with serial relations are not closed under dynamic updates, which can also stand as an issue. Thus, we would like to explore other alternatives in the future, as well as preservation properties on the classes of models.

Another interesting notion we should integrate is the *intentional* aspect of misdirection, i.e., by formalizing which kinds of intention are involved in deception. Here we assumed misdirection to be always intentional and do not investigate further on this aspect. But recent literature covers the intentional dimension from different perspectives, for instance the approach in [49] uses Epistemic Causal Logic to that end. It would be interesting to investigate whether such a setting could accommodate our dynamic logic analysis. Other approaches exist, like [14] (using STIT logic-based framework) and [68] (using causal games) formalizing intentional aspects of misdirection, that we should also take into account for comparison.

Finally, our approach based on dynamic epistemic logic allows us to use tools from *epistemic planning* [10]. As a framework combining standard AI planning with dynamic epistemic logic, epistemic planning could help model strategies of misdirection more accurately. Particularly interesting is the possibility of dealing with agents' (partial) observations. It would be interesting to study the aforementioned task, but where the initial state is an observational epistemic model, and where the action models and the goal(s) are defined over **DLM**, or more specifically from $\mathbf{DLM}^{(+,-)}$. By doing so, we would be able to determine, given some state and a set of actions, if there is an observability-based plan, which leads from the initial state to a (potentially bogus) goal. Another logical tool for reasoning about strategies of misdirection is the family of logics of *knowing how* [66, 67], specially those connected to epistemic planning as [38]. Therein, an agent *knows how to achieve a certain goal* if there exists a sequence of epistemic actions, i.e. an epistemic plan, that transforms the actual model into a model satisfying such a goal. Again, it would be possible to incorporate the action types of **DLM**, or variations of it, into this framework. All these lines of investigation have been briefly discussed, and deserve further exploration in the future.

Acknowledgments. The authors thank two anonymous reviewers for their helpful comments and suggestions, as well as Deborah Marber, Paul Égré, and François Olivier for their feedback on earlier versions of this paper. BI acknowledges the ANRs HYBRINFOX (ANR-21-ASIA-0003) and TRUSTEDNEWS (ANR-25-ASM2-0003), and the programs ECOS-SUD ("Logical consequence and many-valued models", no. A22H01) and THEMIS (grant agreements n°DOS022279400 and n°DOS022279500).

RF is partially supported by projects ANPCyT-PICT-2021-GRF-TI-00400, Stic-AmSud 23-STIC-07 'DL(R)', SecytUNC, the EU Grant Agreement 101008233 (MISSION), and the IRP SINFIN.

References

[1] Thomas Agotnes, Hans P. van Ditmarsch, and Yanjing Wang. True lies. *Synthese*, pages 1–35, 2016.

[2] Alexandru Baltag, Lawrence S. Moss, and Slawomir Solecki. The logic of public announcements and common knowledge and private suspicions. In Itzhak Gilboa, editor, *Proceedings of the 7th Conference on Theoretical Aspects of Rationality and Knowledge (TARK-98)*, pages 43–56, Evanston, IL, USA, 1998. Morgan Kaufmann.

[3] Gaia Belardinelli and Thomas Bolander. Attention! dynamic epistemic logic models of (in) attentive agents. In *Proceedings of the 2023 International Conference on Autonomous Agents and Multiagent Systems*, pages 391–399, 2023.

[4] Gaia Belardinelli and Thomas Bolander. Attention! dynamic epistemic logic models of (in)attentive agents. In Noa Agmon, Bo An, Alessandro Ricci, and William Yeoh, editors, *Proceedings of the 2023 International Conference on Autonomous Agents and Multiagent Systems, AAMAS 2023*, pages 391–399, London, UK, 2023. ACM.

[5] J Bowyer Bell. Toward a theory of deception. *International journal of intelligence and counterintelligence*, 16(2):244–279, 2003.

[6] J Bowyer Bell and Barton Whaley. *Cheating and deception.* Routledge, London, UK, 1991.

[7] P. Blackburn, M. de Rijke, and Y. Venema. *Modal Logic.* Cambridge University Press, UK, 2001.

[8] Patrick Blackburn, Johan F. A. K. van Benthem, and Frank Wolter. *Handbook of Modal Logic*, volume 3. Elsevier Science Inc., New York, NY, USA, 2006.

[9] Sissela Bok. *Secrets: On the ethics of concealment and revelation.* Oxford University Press, New York, USA, 1983.

[10] Thomas Bolander, Tristan Charrier, Sophie Pinchinat, and François Schwarzentruber. DEL-based epistemic planning: Decidability and complexity. *Artif. Intell.*, 287:103304, 2020.

[11] Thomas Bolander, Hans van Ditmarsch, Andreas Herzig, Emiliano Lorini, Pere Pardo, and François Schwarzentruber. Announcements to attentive agents. *Journal of Logic, Language and information*, 25(1):1–35, 2016.

[12] Denis Bonnay and Paul Égré. Vagueness and Introspection. April 2007.

[13] Denis Bonnay and Paul Égré. Inexact knowledge with introspection. *Journal of Philosophical Logic*, 38(2):179–227, 2009.

[14] Grégory Bonnet, Christopher Leturc, Emiliano Lorini, and Giovanni Sartor. Influencing choices by changing beliefs: A logical theory of influence, persuasion, and deception. In Stefan Sarkadi, Benjamin Wright, Peta Masters, and Peter McBurney, editors, *Deceptive AI*, pages 124–141, Cham, 2021. Springer International Publishing.

[15] Thomas L Carson. *Lying and deception: Theory and practice.* OUP Oxford, Oxford, UK, 2010.

[16] Tristan Charrier, Andreas Herzig, Emiliano Lorini, Faustine Maffre, and François Schwarzentruber. Building epistemic logic from observations and public announcements. In *15th International Conference on the Principles of Knowledge Representation and Reasoning*, 2016.

[17] Brian F. Chellas. *Modal Logic: An Introduction.* Cambridge University Press, New York, USA, 1980.

[18] Roderick M. Chisholm and Thomas D. Feehan. The intent to deceive. *Journal of Philosophy*, 74(3):143–159, 1977.

[19] Erhan Demircioglu. Dretske on non-epistemic seeing. *Theoria*, 83, 06 2017.

[20] Fred I Dretske. Seeing and knowing. *British Journal for the Philosophy of Science*, 21(1), 1970.

[21] Fred I. Dretske. Simple seeing. In Donald F. Gustafson and Bangs L. Tapscott, editors, *Body, Mind, and Method*, pages 1–15. Kluwer Academic Publishers, Alphen aan den Rijn, NL, 1979.

[22] Paul Égré. Half-truths and the liar. In Johannes Stern and Carlo Nicolai, editors, *Modes of Truth. The Unified Approach to truth, modality and paradox.* Routledge, London, UK, 2021.

[23] Paul Égré and Benjamin Icard. Lying and vagueness. In Jörg Meibauer, editor, *Oxford Handbook of Lying.* Oxford University Press, Oxford, UK, 2018.

[24] Don Fallis. A conceptual analysis of disinformation. *iConference 2009*, 2009.

[25] Don Fallis. Shedding light on keeping people in the dark. *Topics in Cognitive Science*, 12(2):535–554, 2020.

[26] Jie Fan, Yanjing Wang, and Hans Van Ditmarsch. Contingency and knowing whether. *Rev. Symb. Log.*, 8(1):75–107, 2015.

[27] David Fisher. *The War Magician: The man who conjured victory in the desert.* Hachette UK, London, UK, 2011.

[28] Dariel Fitzkee. *Magic by misdirection.* Ravenio Books, New York, USA, 1975.

[29] Peter Forbes. *Dazzled and deceived: mimicry and camouflage.* Yale University Press, New Haven, USA, 2011.

[30] Karl Fulves. *Big Book of Magic Tricks.* Dover Publications, New York, USA, 1977.

[31] Andreas Herzig, Emiliano Lorini, and Faustine Maffre. A poor man's epistemic logic based on propositional assignment and higher-order observation. In *International Workshop on Logic, Rationality and Interaction*, pages 156–168. Springer, 2015.

[32] Jaakko Hintikka. *Knowledge and Belief.* Cornell University Press, Ithaca N.Y., 1962.

[33] Wesley H. Holliday and Thomas F. Icard III. Moorean phenomena in epistemic logic. In Lev D. Beklemishev, Valentin Goranko, and Valentin B. Shehtman, editors, *Advances in Modal Logic 8*, pages 178–199, Moscow, Russia, 2010. College Publications.

[34] Ekaterina Kubyshkina and Mattia Petrolo. A logic for factive ignorance. *Synthese*, 198(6):5917–5928, 2021.

[35] Gustav Kuhn, Hugo A Caffaratti, Robert Teszka, and Ronald A Rensink. A psychologically-based taxonomy of misdirection. *Frontiers in psychology*, 5:1392, 2014.

[36] Gustav Kuhn, Jay A Olson, and Amir Raz. The psychology of magic and the magic of psychology. *Frontiers in Psychology*, 7:1358, 2016.

[37] Fenrong Liu. Diversity of agents and their interaction. *Journal of Logic, Language and information*, 18(1):23–53, 2009.

[38] Fenrong Liu and Yanjing Wang. Reasoning about agent types and the hardest logic puzzle ever. *Minds and Machines*, 23(1):123–161, 2013.

[39] Emiliano Lorini and Cristiano Castelfranchi. The unexpected aspects of surprise. *International Journal of Pattern Recognition and Artificial Intelligence*, 20(06):817–833, 2006.

[40] Emiliano Lorini and Cristiano Castelfranchi. The cognitive structure of surprise: looking for basic principles. *Topoi*, 26:133–149, 2007.

[41] James E Mahon. The definition of lying and deception. *The Stanford Encyclopedia of Philosophy*, 2015.

[42] Jasper Maskelyne. *Magic–top Secret.* London, New York, S. Paul, London, UK, 1949.

[43] John Nevil Maskelyne and David Devant. *Our magic.* Munro's Modern Magical Manufacturers, London, UK, 1911.

[44] Wulf-Uwe Meyer, Rainer Reisenzein, and Achim Schützwohl. Toward a process analysis of emotions: The case of surprise. *Motivation and Emotion*, 21(3):251–274, 1997.

[45] Andrew Ortony and Derek Partridge. Surprisingness and expectation failure: what's the difference? In *IJCAI*, pages 106–108, 1987.

[46] Donald LJ Quicke. *Mimicry, crypsis, masquerade and other adaptive resemblances.* John Wiley & Sons, Hoboken, USA, 2017.

[47] Graeme D Ruxton, William L Allen, Thomas N Sherratt, and Michael P Speed. *Avoiding attack: the evolutionary ecology of crypsis, aposematism, and mimicry.* Oxford university press, Oxford, UK, 2019.

[48] Chiaki Sakama. A formal account of deception. In *Proceedings of the AAAI Fall 2015 Symposium on Deceptive and Counter-Deceptive Machines*, pages 34–41, 2015.

[49] Chiaki Sakama. Deception in epistemic causal logic. In Stefan Sarkadi, Benjamin Wright, Peta Masters, and Peter McBurney, editors, *Deceptive AI*, pages 105–123, Cham, 2021. Springer.

[50] Chiaki Sakama and Martin Caminada. The many faces of deception. *Proceedings of the Thirty Years of Nonmonotonic Reasoning*, 2010.

[51] Chiaki Sakama, Martin Caminada, and Andreas Herzig. A logical account of lying. In Tomi Janhunen and Ilkka Niemelä, editors, *Logics in Artificial Intelligence – 12th European Conference, JELIA 2010, Helsinki, Finland, September 13–15, 2010, Proceedings*, volume 6341 of *Lecture Notes in Computer Science*, pages 286–299. Springer, 2010.

[52] Chiaki Sakama, Martin Caminada, and Andreas Herzig. A formal account of dishonesty. *Logic Journal of the IGPL*, 23(2):259–294, 2014.

[53] Chiaki Sakama and Katsumi Inoue. Abduction, conversational implicature and misleading in human dialogues. *Logic Journal of IGPL*, 24:526—541, 2016.

[54] S. H. Sharpe. *Conjurers' Psychological Secrets.* Hades Publications, 1988.

[55] Samuel H Sharpe. *Neo-Magic: The Art of the Conjurer.* George Johnson Magical Publications, London, UK, 1932.

[56] AD Smith. Perception and belief. *Philosophy and phenomenological research*, 62(2):283–309, 2001.

[57] Wally Smith, Frank Dignum, and Liz Sonenberg. The construction of impossibility: a logic-based analysis of conjuring tricks. *Frontiers in psychology*, 7:748, 2016.

[58] Dirk van Dalen. *Logic and Structure.* Universitext. Springer London, New York, USA, 2012.

[59] Wiebe van der Hoek and Alessio Lomuscio. A logic for ignorance. *Electronic Notes in Theoretical Computer Science*, 85(2):117 – 133, 2004. LCMAS 2003, Logic and Communication in Multi-Agent Systems.

[60] Hans van Ditmarsch, Andreas Herzig, and Tiago de Lima. Public announcements, public assignments and the complexity of their logic. *Journal of Applied Non Classiscal Logics*, 22(3):249–273, 2012.

[61] Hans van Ditmarsch and Barteld Kooi. Semantic results for ontic and epistemic change. In *Logic and the Foundations of Game and Decision Theory (LOFT 7)*, pages 87–118. Amsterdam University Press, NL, 2008.

[62] Hans P. van Ditmarsch. Dynamics of lying. *Synthese*, 191(5):745–777, 2014.

[63] Hans P. van Ditmarsch, Jan van Eijck, Floor Sietsma, and Yanjing Wang. On the logic of lying. In Jan van Eijck and Rineke Verbrugge, editors, *Games, Actions and Social Software: Multidisciplinary Aspects*, pages 41–72. Springer Berlin Heidelberg, Wassenaar, NL, 2012.

[64] Hans van Van Ditmarsch, Wiebe van Der Hoek, and Barteld Kooi. *Dynamic epistemic logic*, volume 337. Springer, Dordrecht, NL, 2007.

[65] George H. Von Wright. *An Essay in Modal Logic*. North Holland, Amsterdam, 1951.

[66] Yanjing Wang. A logic of knowing how. In *Logic, Rationality, and Interaction – 5th International Workshop, LORI 2015*, pages 392–405, Taipei, Taiwan, 2015.

[67] Yanjing Wang. A logic of goal-directed knowing how. *Synthese*, 195(10):4419–4439, 2018.

[68] Francis Rhys Ward, Francesca Toni, and Francesco Belardinelli. Defining deception in structural causal games. In Noa Agmon, Bo An, Alessandro Ricci, and William Yeoh, editors, *Proceedings of the 2023 International Conference on Autonomous Agents and Multiagent Systems, AAMAS 2023*, pages 2902–2904, Richland, SC, USA, 2023. ACM.

[69] Barton Whaley. Stratagem: deception and surprise in war. Technical report, Cambridge, Mass.: Center for International Studies, Mass. Institute, USA, 1969.

Received July 2025

Intra and Inter-paradigm Reconfigurations with ZX: An Exercise

Bruno Jardim
HASLab INESC TEC & Universidade do Minho,
Campus de Gualtar, 4710-057 Braga, Portugal
bruno.f.jardim@inesctec.pt

Jaime Santos
HASLab INESC TEC & Universidade do Minho,
Campus de Gualtar, 4710-057 Braga, Portugal

Luís S. Barbosa
HASLab INESC TEC & Universidade do Minho,
Campus de Gualtar, 4710-057 Braga, Portugal

Abstract

Current implementations of quantum walks within the measurement based paradigm are limited to implementations specifically designed for that paradigm. Thus, they bear very little correlation with the most common implementations of quantum walks within the circuit model. This encourages the establishment of a direct comparison for these quantum protocols.

This paper shows that, through a graph reconfiguration process, it is possible not only to generate an alternative evolution operator for staggered quantum walks within the circuit model (*intra*-paradigm reconfiguration), but also to derive its implementation in a completely different paradigm — the measurement based paradigm (*inter*-paradigm reconfiguration).

This work was supported by FCT, the Portuguese Foundation for Science and Technology, within the project BANSKY, with reference COMPETE2030-FEDER-00892000, a FCT PhD grant for the first author (2025.04550.BD).

1 Introduction

A reconfiguration process requires a form of representation of its object (e.g. a computational system, a program, a communication protocol) and a set of rules that govern the possible transformation between different forms such a representation may acquire. The sort of transformations to be considered vary with with domain of application and/or the associated engineering methodology. For example, they can entirely preserve meaning, making the system evolve through isomorphic configurations, or preserve instead a given subset of properties, typically related to safety or liveness requirements.

A set of reconfiguration rules becomes a calculus if proved complete for a given goal: in such a case rule application always leads to valid configurations and all such configurations can indeed be obtained that way. Finally, a calculus gives rise to a dynamic reconfiguration engine if it can be fully automated and applied on the fly.

This paper discusses reconfiguration in a very specific setting: objects are quantum programs and the reconfiguration process resorts to a graphical calculus — the ZX-calculus [1, 6] — which provides a complete rewriting system for such programs. Being equipped with a kernel of tools that perform systematically rule application, the whole process becomes automated. Moreover, as quantum programs are represented as (specific sorts of) graphs, what is basically discussed in the paper can be framed within the broader domain of graph reconfiguration.

This work extends our previous conference paper published in the Proceedings of ReacTS'2024 [3]. That paper introduced an optimization exercise targeting a very general variant of discrete-time quantum walks — the *staggered* model [7, 8].

Quantum walks [13] are the quantum equivalent to random walks in which, differently from the classical case, where the walker's next move follows the result of some sort of random choice, evolution typically proceeds in an equally weighed superposition of possible moves through the iteration of a unitary operator, without resorting to intermediate measurements. The staggered model, avoiding the use of a coin to direct the walker evolution, explores the underlying graph structure to build an evolution operator based on local unitaries induced by adjacent vertices.

Any quantum protocol, no matter the paradigm in which it is defined, can be abstracted through the ZX-calculus in the form of a ZX-diagram, i.e. a bi-coloured (red/green) graph. This was used in our previous work as a way to achieve an alternative evolution operator for the *staggered quantum walk*. Reconfiguration was circumscribed, in that exercise revisited in section 3 below, to the usual circuit model of quantum computing.

This paper, however, takes a more ambitious path. Rather than simply resort to the ZX-calculus to obtain an optimized version of the *staggered quantum walk* circuit, working always within the well-known circuit model, we investigate the calculus to sustain a *graph reconfiguration* process that yields a version of this protocol within an entirely different paradigm, the so-called *measurement-based quantum computing* (MBQC) paradigm [10]. Therefore, the focus of the reconfiguration is no more circuit optimization, but a deeper representation change, referred in the sequel as a *inter-paradigm reconfiguration.*

The paper is organised as follows. Section 2 briefly introduces the ZX-calculus and illustrates how a simple quantum circuit is translated and rewritten in ZX, a process supported by the *PyZX* tool [4]. Then, sections 3 and 4 discuss both *intra-* and *inter-*paradigm reconfiguration in the context of a concrete exercise with implementations of the staggered quantum walk circuit. Section 3 starts from the basic quantum circuit for this sort of quantum walks and carries on a graph reconfiguration process towards its optimization, along the lines of [3]. Section 4 motivates *inter*-paradigm reconfiguration and sets the scene for what follows introducing the measurement-based paradigm for quantum computing as an increasingly popular alternative to the circuit model. Section 5 carries out in detail the envisaged exercise reconfiguring the circuit model version of the staggered quantum walk protocol in the new, measurement-based paradigm. Finally, section 6 concludes the paper and identifies some topics for future work.

2 A Brief Introdution to ZX Diagrams and Calculus

The ZX-calculus is a graphical language and a set of graph transformation rules for reasoning about quantum programs. As mentioned in the Introduction, the calculus is both both *sound* and *complete.* Therefore, any equality between linear maps can be proven by graph rewriting.

ZX-diagrams, which boil down to graphical representations of linear maps between quantum states, are composed of two distinct node types, referred to as *spiders*, typically coloured green and red, and connected through wires. Green and red, spiders represent rotations around the Z- and X-axis in the Bloch sphere representation of a qubit, respectively. A real number (in the following diagrams represented by α), ranging between $[0, 2\pi)$ represents the phase of any specific spider, and is generally omitted when zero.

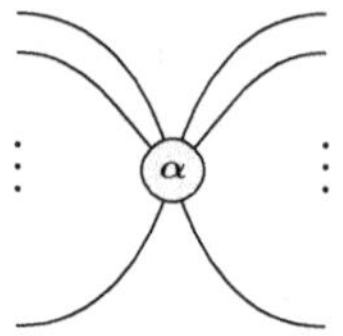

$$:= \quad |0\cdots 0\rangle\langle 0\cdots 0| + e^{i\alpha}|1\cdots 1\rangle\langle 1\cdots 1|$$

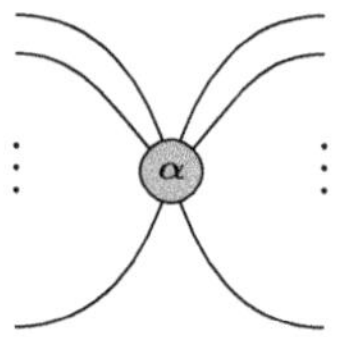

$$:= \quad |+\cdots+\rangle\langle+\cdots+| + e^{i\alpha}|-\cdots-\rangle\langle-\cdots-|$$

Along with these generators there is another graphical element that appears in the diagrams besides the spiders and wires: the yellow box. This stands for a representation of the ubiquitous Hadamard gate in quantum computing. Indeed, the gate can be represented with Z- and X-spiders, through its Euler angle decomposition, which sometimes is simplified even further as a dashed blue wire. The following small example illustrates how a circuit is translated to a ZX-diagram:

The calculus introduces a set of rewrite rules, which embody two overarching principles:

- Only connectivity matters.

- Everything remains equal if red and green nodes are swapped.

The former allows us to distort a diagram as much as we may please, as long as we keep the same connections. This principle will be crucial in section 5 to establish inter-paradigm equivalence.

The latter rule embodies, in a certain sense, the principle of duality. That is, if a certain derivation is true for a node, or set of nodes of a given color, it will also hold for the opposite color.

The following rewrite rules, combined with a set of diagram equivalences (up to a scalar), materializes a calculus which is also *universal*, in the sense that any linear map between qubits can be represented as a ZX-diagram. This also implies that any quantum circuit has a ZX counterpart. However, the opposite might not be guaranteed, due to the additional expressiveness of the ZX-calculus with respect to the quantum circuit model.

Rewrite Rules

- Spider Fusion (sf)

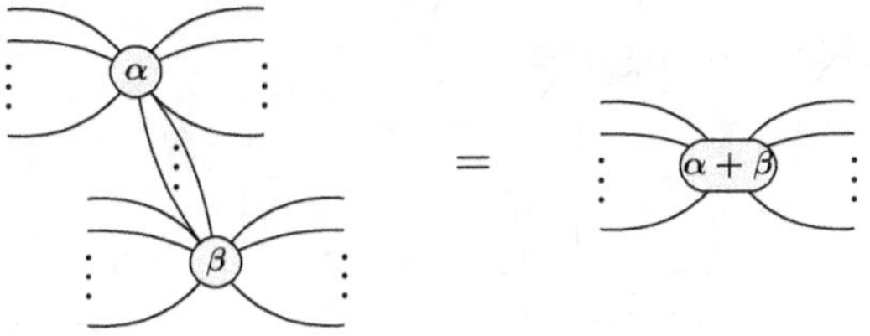

(1)

- Identity Removal (id)

(2)

- Hadamard cancellation (id)

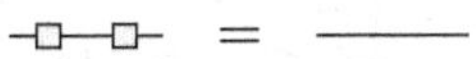

(3)

- π commutation (π)

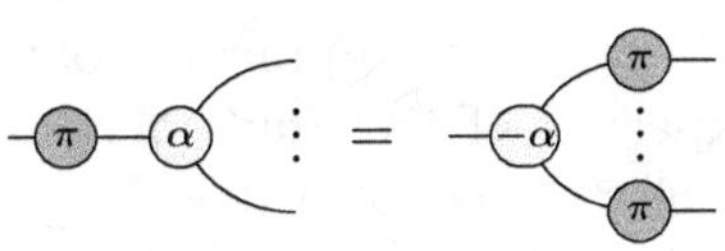

(4)

- State copy (sc)

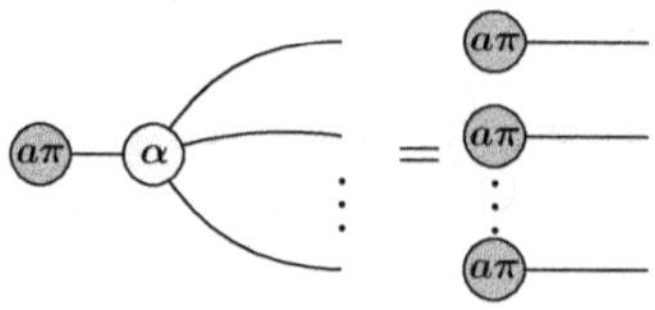

$a \in \{0, 1\}$ (5)

- Color change (cc)

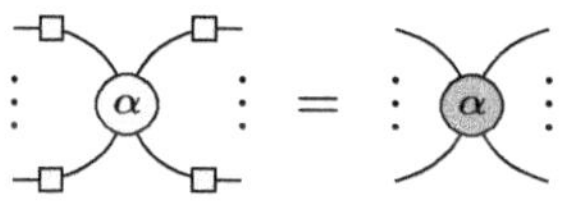

(6)

- Strong complementarity (bg)

(7)

- Hopf (hopf)

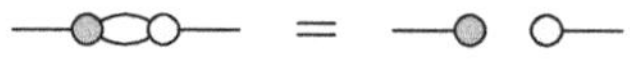

(8)

The following is a ZX-calculus graphical proof that a circuit that creates the GHZ-state behaves as expected. This simple derivation provides an example to the expressiveness and simplicity of the ZX-calculus.

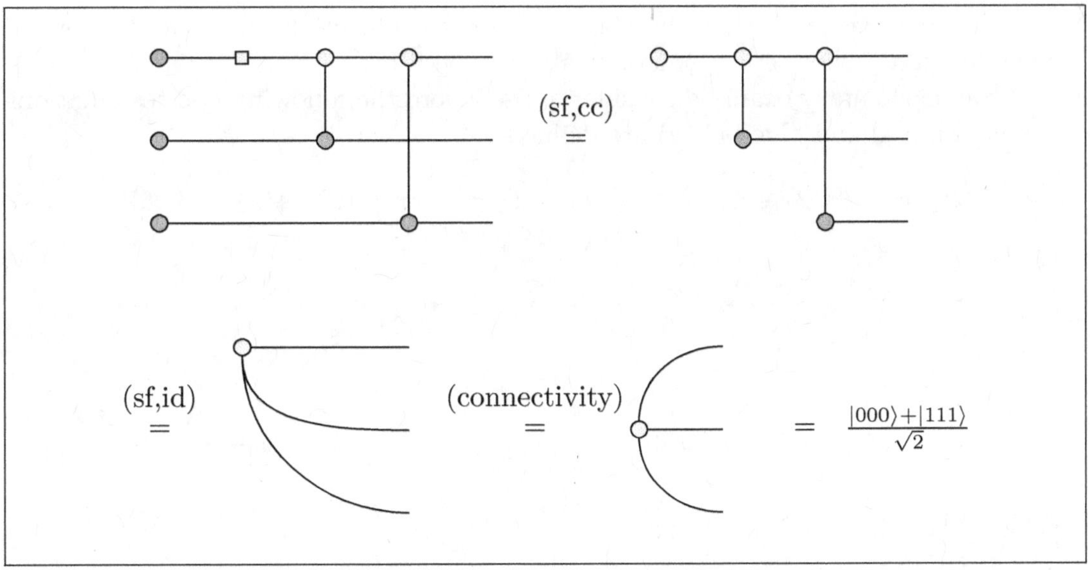

3 Intra-paradigm Reconfiguration: An Optimization of the Staggered Quantum Walk Circuit

3.1 The Staggered Quantum Walk

Staggered walks [8] explore partitions of graph cliques (subsets of vertices in which every two distinct vertices are adjacent) over the graph structure of the walking space. Each partition forms a tessellation whose elements do not overlap. The set of cliques in each tesselllation must cover all vertices of the graph, and the set of tesselllations $\{T_1, T_2, \ldots, T_k\}$ chosen must cover all the edges. Then a unit vector, typically encoding a uniform superposition, is associated to each clique so that the vector belongs to the subspace spanned by the corresponding vertices; i. e., $\left|u_j^k\right\rangle = \frac{1}{\sqrt{|\alpha_j^k|}} \sum_{l \in \alpha_j^k} |l\rangle$, where α_j^k is the jth polygon in the kth tesselllation. This way each tesselllation k gives rise to an operator $H_k = 2\sum_{j=1}^{p} \left|u_j^k\right\rangle \left\langle u_j^k\right| - I$ which propagates the probability amplitude locally, in each clique. The composition of all such operators defines the evolution operator, which, by solving the the time-independent Schrödinger equation, is equivalent to

$$U = e^{i\theta_k H_k} \ldots e^{i\theta_2 H_2} e^{i\theta_1 H_1}, \quad \text{where} \quad e^{i\theta_k H_k} = \cos(\theta_k) I + i \sin(\theta_k) H_k$$

since $H_k^2 = I$, meaning that the Hamiltonian is a reflection operator that, when expanded in a Taylor series, generates a local operator.

As an elementary example consider a line where the following two tesselllations (depicted in red and blue below) are defined

$$T_\alpha = \{\{2x, 2x+1\} \colon x \in \mathbb{Z}\} \text{ and } T_\beta = \{\{2x+1, 2x+2\} \colon x \in \mathbb{Z}\}.$$

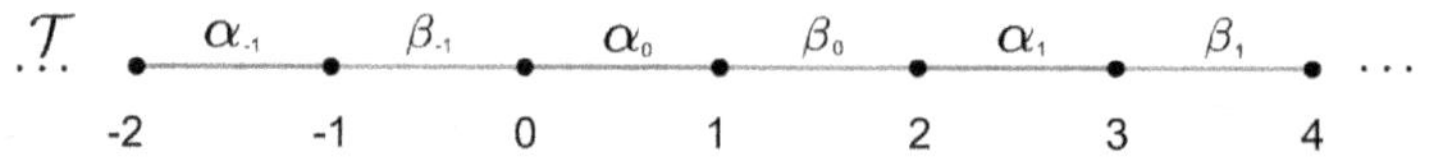

Thus,

$$|\alpha_x\rangle = \frac{|2x\rangle + |2x+1\rangle}{\sqrt{2}} \text{ and } |\beta_x\rangle = \frac{|2x+1\rangle + |2x+2\rangle}{\sqrt{2}},$$

yielding the following Hamiltonians

$$H_\alpha = 2 \sum_{x=-\infty}^{+\infty} |\alpha_x\rangle \langle \alpha_x| - I \text{ and } H_\beta = 2 \sum_{x=-\infty}^{+\infty} |\beta_x\rangle \langle \beta_x| - I.$$

Therefore, $U = e^{i\theta H_\beta} e^{i\theta H_\alpha}$ is the evolution operator. The probability distribution on a line after 50 steps, starting at $|+\rangle$, for different values of θ, is depicted Fig. 1 noticing that the walker is more likely to be found further away from the origin as the angle increases.

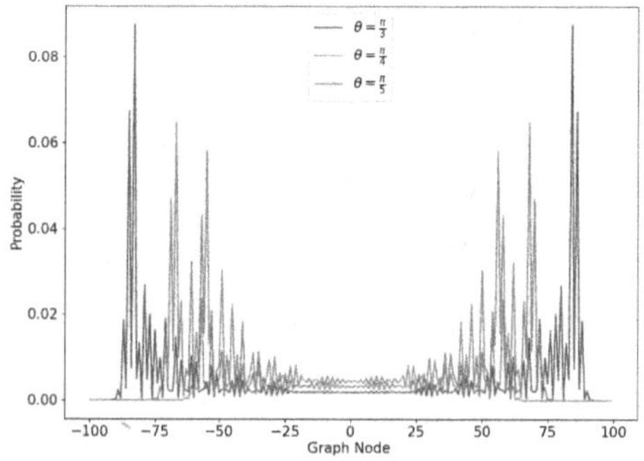

Figure 1: Probability distribution on a line dependent on θ, 50 steps, from $|+\rangle$.

3.2 Circuit Optimization

A circuit implementation of the staggered model can be found in [11]. For the example discussed above, it yields the circuit represented in Fig. 2 where $R_x(\theta) = e^{\frac{-i\theta X}{2}}$ and the Inc(rement) and Dec(rement) circuits have the usual implementation through generalised Toffoli gates. When the walker reaches the limit of the state space it cycles back.

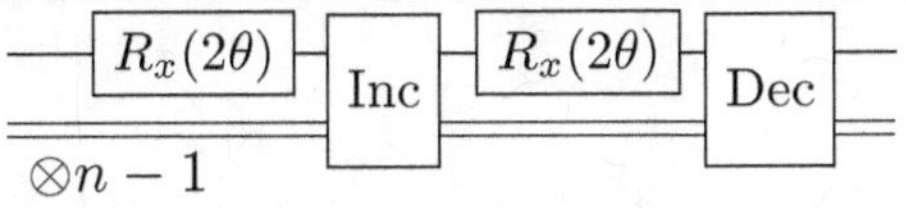

Figure 2: A circuit implementation of the staggered model [11].

An implementation of a 3 qubit staggered quantum walk, taking $\theta = \frac{\pi}{3}$, which maximizes[1] propagation, is represented in the ZX diagram depicted in Fig. 3 using ZH-calculus H-box notation for a concise representation of Toffoli gates.

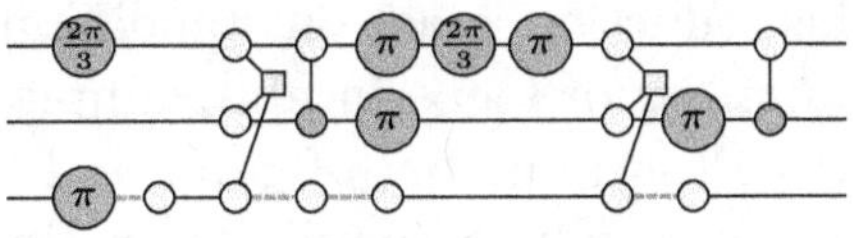

Figure 3: A 3 qubit staggered quantum walk in ZX.

Advanced techniques, described in [5] and directly implemented in *PyZX* [4] as the `full_reduce` method, may reduce the circuit T-count in about 50% [5]. Although this is not the case for our small example, when we start applying such simplifications to staggered models with larger amounts of steps the T-count reduction can reach approximately 60-70%. Back to the example, this simplification yields the graph depicted in Fig. 4

[1]For this specific graph $\theta = \frac{\pi}{3}$ maximizes propagation, however $\frac{\pi}{2}$ is the optimal parameter for a complete graph.

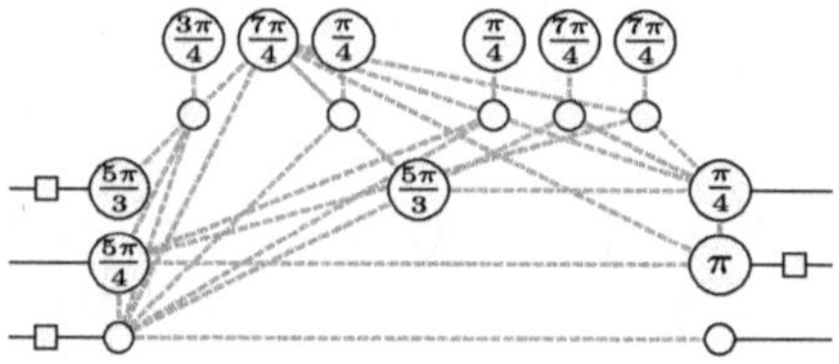

Figure 4: The resulting ZX-diagram after a `full_reduce` optimization routine.

This diagram no longer resembles a circuit, making a comparison with the original one difficult. The circuit extracted [2] by *PyZX* has more gates than the one obtained from simple manual optimizations, although the T-count is indeed smaller. In fact, the `full_reduce` method introduces several additional Hadamard gates. The subsequent circuit extraction 'preserves' the nature of the *graph-like* ZX-diagram. Following the extraction with a small set of simplifications, basically resorting to the fusion rewrite rule followed by a color-change, we get a much smaller circuit. This fully-simplified circuit now surpasses the original one in both the total amount of gates and T-count. Although this reduction is not outstanding in this example, it becomes most relevant when the number of steps in an example increases. The following tables show, respectively, the total number of gates and the T-count value induced by the different optimization procedures for the same 3-qubit implementation.

	Number of steps in the staggered quantum walk			
Optimizations:	1 step	2 steps	4 steps	8 steps
None	39	77	153	305
Full-reduce + fusion/id/to_rg	37	47	72	118

	Number of steps in the staggered quantum walk			
Optimizations:	1 step	2 steps	4 steps	8 steps
None	16	32	64	128
Full-reduce + fusion/id/to_rg	10	16	28	52

3.3 An Alternative Evolution Operator

When analysing the ZX diagram for a long staggered quantum walk (i.e. with more than 5 steps) a pattern starts to emerge, repeating itself as many times as the number of steps considered. Depicted in ZX in Fig. 5, it seems able to approximate, both the increment and decrement layers of the evolution operator.

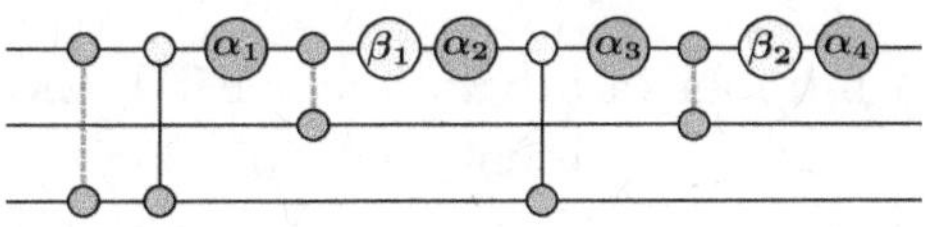

Figure 5: The pattern emerging through reconfiguration.

where $\alpha_n = \pm\frac{\pi}{4}$ and $\beta_n = \frac{2\pi}{3} + m\pi$, with $m = 0$ or $m = 1$. There is also a slight variation of this operator, where a CNOT gate between the first and last qubit appears right after the β_1 Z-spider. This diagram does not fully capture the staggered model we started with, but, once suitably enveloped, it captures the exact same tensor as the original circuit. The set of gates to be placed as an envelope, in the beginning and the end of the diagram, does not exhibit a specific structure. This construction appeared when optimising the 3 qubit staggered quantum walk. However, it can be generalised for an n qubit implementation, yielding the following operator, in the form of a ZX-diagram as depicted in Fig. 6.

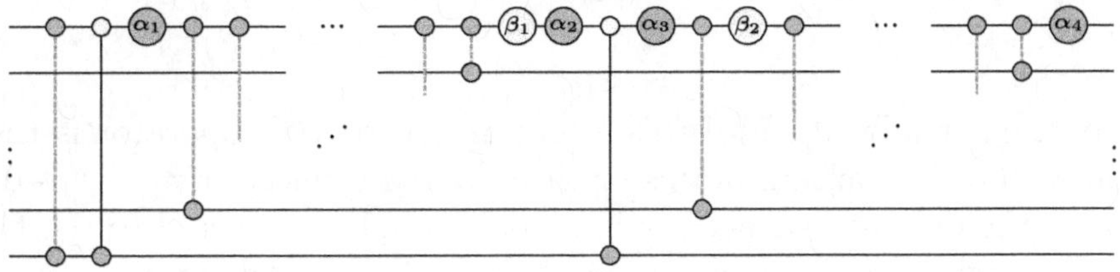

Figure 6: The n-qubit case.

The rationale behind this operator is easy to explain: it creates a uniform distribution over a certain number of states, applies a rotation that makes some states more likely than others and then spreads these probabilities over the remaining states using CNOT gates. This also explains why the pattern only shows up in staggered quantum walks over a certain length. The classical evolution operator needs to be repeated a number of times to be able to spread the probability distributions over the whole state space. This is exactly what this version does on the first layer.

In any case, it has a number of advantages. First and foremost it reduces the total amount of gates needed to represent the evolution of the quantum walk. With the number of qubits increasing so does the cost of the increment and decrement layers, as a n qubit staggered quantum walk needs to implement MCX gates with $n-1$ controls. The alternative operator derived here uses gates controlled by at most one qubit. Moreover, to go from a n to a $n+1$-qubit quantum walk, all that needs to be done is to add two more XCX-gates, one to each ladder of XCX-gates.

In general, this makes the alternative operator much more efficient with respect to the total number of gates used, leading to lower depth and, therefore, potentially less error-prone circuits.

As mentioned above, just by itself this operator can approximate the evolution of a staggered quantum walk. Although the approximation is not perfect it can yield results which are quite similar to the ones obtained with the original implementation of the staggered model, as shown in the graph in Fig. 7, where the original operator is represented in red and the alternative in blue.

One particular advantage of this alternative evolution operator is that it can work quite well on a quantum processor with limited connectivity. This is due to the fact that all the qubits used need only to be strongly connected to the first qubit. However, a number of challenges remain, requiring further investigation. These concern the most suitable choice of parameters for α_n and β_n, as well as whether and how they depend on the number of qubits used in a particular staggered walk. Actually, when optimising the four qubits implementation of this circuit the resulting parameters did not seem to follow any regular pattern.

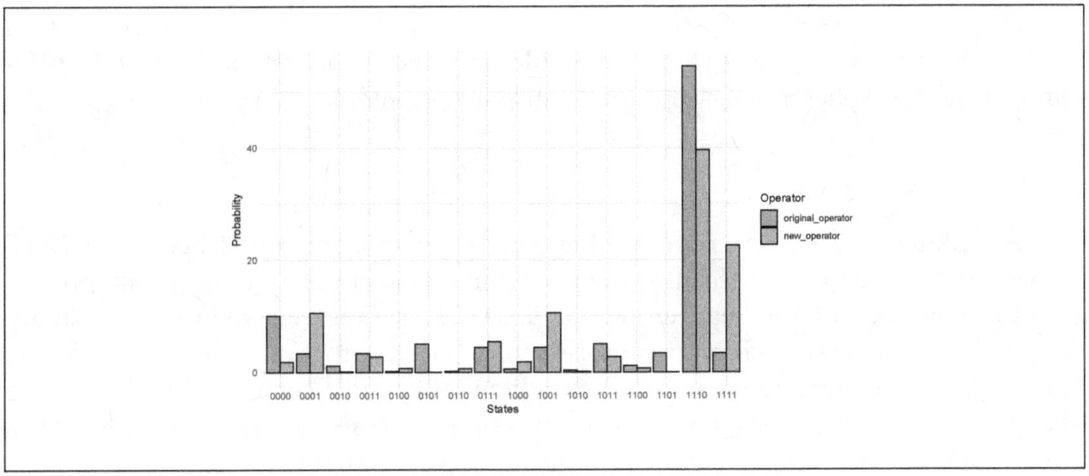

Figure 7: Comparing both operators.

4 Inter-paradigm Reconfiguration: Towards a MBQC Implementation

In the previous section, ZX rewrite rules were used to optimize a circuit implementation of staggered quantum walks. The proposed optimization was not specific to a given metric, but instead represented a balance between the total amount of gates and the T-count. Both metrics are essential to achieve decoherence resistant quantum protocols. The total amount of gates is deeply linked to the depth of the circuit: as it grows more likely the circuit becomes to be affected by decoherence. On the other hand, the reduction of the T-count is significant when taken into consideration the cost of implementing the T-gates.

From the optimization process described, a pattern emerged for quantum walks that spanned for four or more steps. This pattern was contained between a set of, seemingly random, gates at the beginning and end of the circuit, and it repeats itself as many times as the steps in the quantum walk. This suggests some sort of correlation between the evolution dynamics of the staggered quantum walk and this pattern. Indeed, when suitably parameterized, it acts as good approximation of the former, as expected.

In comparison with the standard implementation that was taken as our starting point (see Fig. 3), this alternative evolution operator exhibits much less depth for a single step and uses a set of gates that is much cheaper to implement. This, however, comes at the cost of varying precision when approximating a given staggered quantum walk. Nonetheless it proves to be a suitable implementation to more limited QPUs and/or more decoherence susceptible quantum computers.

A striking aspect of the optimized circuit is the absence of multi-control CNOT gates. As a matter of fact this is quite uncommon as such gates are typically present in any implementation of quantum walks. This observation triggered the opportunity to proceed with the reconfiguration process towards a different quantum programming paradigm — the measurement based paradigm [10]. Multi-control CNOT gates are based on a quite expensive decomposition process into smaller gates, an effort that becomes unmanageable in the measurement-based paradigm.

Before discussing in the next section how a measurement-based implementation of the staggered quantum walk is obtained, let us briefly describe this paradigm and highlight its challenges.

Measurement Based Quantum Computing (MBQC) is an alternative paradigm for quantum computing which relies only on measurements on a cluster state to advance the computational process. Measurements are performed on specific nodes of the state and then propagated. The paradigm has gained a lot of traction over the last few years, mostly due to suitable applications in photonic quantum platforms which are relatively low-cost compared to other architectures, and much less prone to decoherence.

MBQC entirely relies on single-qubit measurements over a heavily entangled initial state, usually denoted by the cluster state. These measurements are a destructive and inherently random process. However, acting over highly entangled states they dictate the evolution of the program. The processing of measurement outcomes in a specific order and in a specific basis allows for their propagation through the cluster. The cluster state is created by entangling pairs of qubits in a grid-like fashion with CZ-gates, which can be represented in the ZX-diagram depicted in Fig. 8.

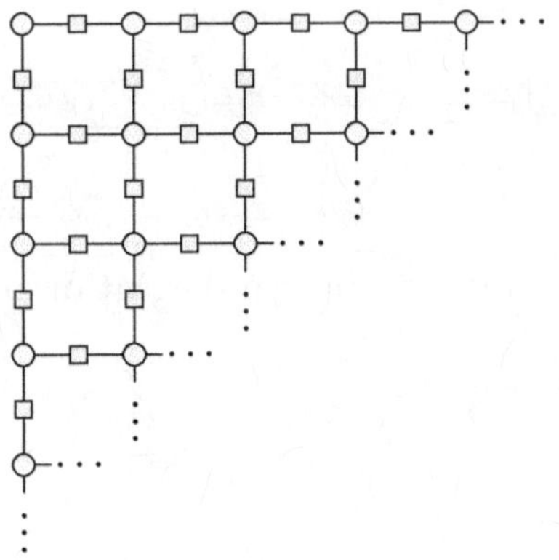

(9)

Figure 8: The cluster state represented as a ZX-diagram.

The number of qubits in each row can be regarded as the number of qubits available to perform computations. We also assume that the information flows from the left to the right. Note that the cluster state does not need to be fully constructed before the algorithm is set in motion. It can be constructed as the computation goes on, as long as the building process is faster than destruction by measurement. This makes possible the reuse of qubits and thus allows for processes that are much larger than the initial cluster state upon which they are executed.

5 Inter-paradigm reconfiguration: The exercise

We are now ready to explore the reconfiguration of the optimized circuit for staggered quantum walks into the measurement-based paradigm. Our starting point are the following two reconfigurations introduced in [12] and depicted in figures 9 and 10. They provide the basis to transform unitary gates into a corresponding sequence of measurements and corrections.

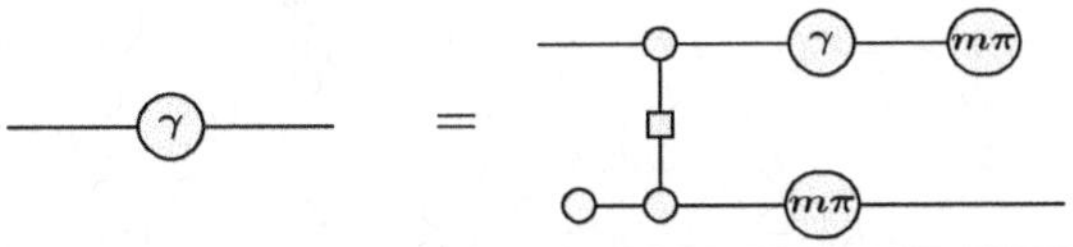

Figure 9: The inter-paradigm derivation of a ZRot-gate.

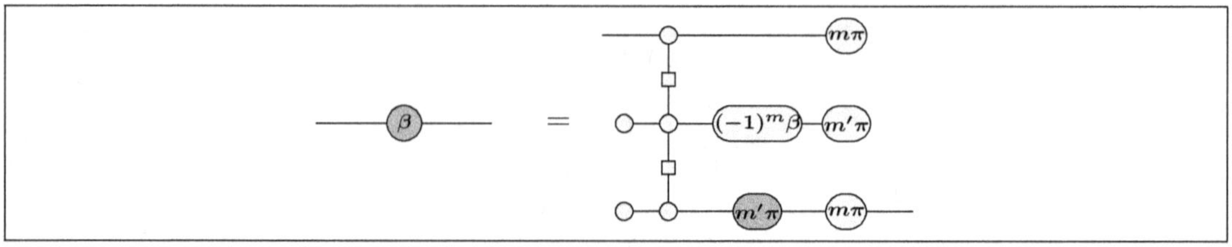

Figure 10: The inter-paradigm derivation of a XRot-gate .

Implementing staggered quantum walks within the framework of MBQC is quite challenging, basically due to the need to eliminate all multi-control CNOT gates. The alternative, optimized circuit, discussed in section 3, however, has already eliminated most source of difficulty, relying only on simple CNOT, Hadamard and X- and Z-rotation gates.

Therefore, the necessary step involves the reconfiguration of X- and Z-rotations, as described above, as well as, in a similar manner, XCX and CNOT gates. This is achieved by the *verticalization* of these gates, separating their inputs from their outputs, as specified in the two rules in Fig. 11 and Fig. 12, respectively. The positions of the inputs and output are interchangeable, which is essential to guarantee the re-usability of the cluster state.

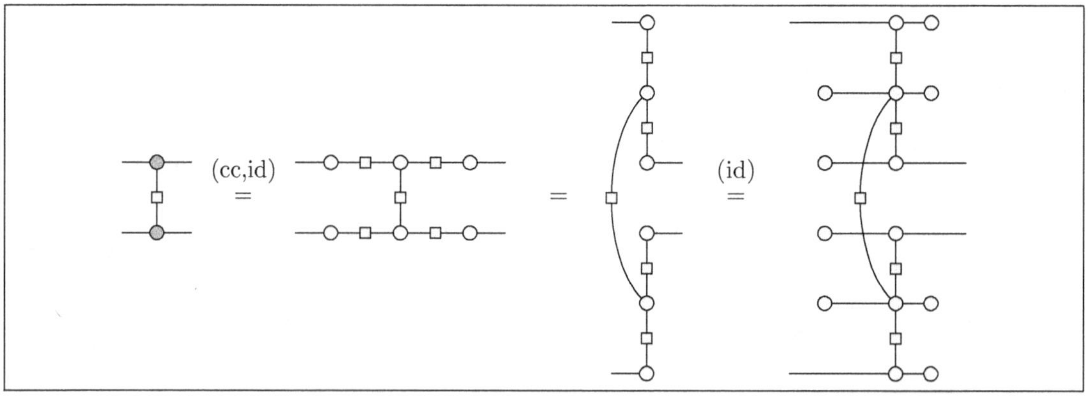

Figure 11: The inter-paradigm derivation of a XCX-gate.

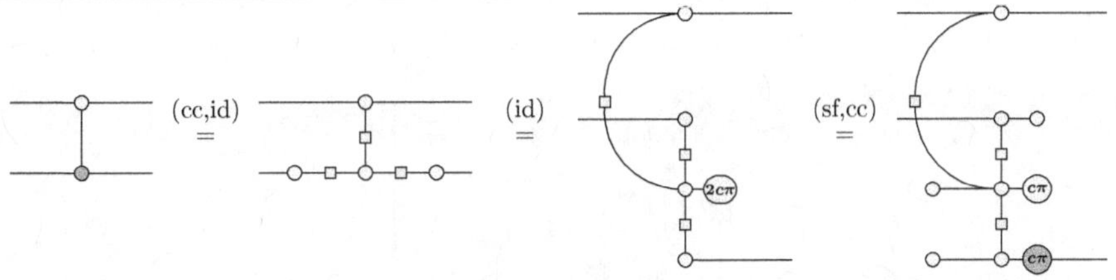

Figure 12: The inter-paradigm derivation of a CNOT-gate.

For the CNOT case, as represented in Fig. 12, a measurement was necessary to suitably correct the target qubit.

Recall now our initial circuit, i.e. the staggered quantum walk after the intra-reconfiguration process discussed in section 3. As shown in [3] the operator it encodes, if parameterized correctly, is able to approximate the evolution of the staggered quantum walk if the latter has a length greater than four steps.

By replacing the gates in the circuit with their MBQC counterparts, according to the reconfiguration rules introduced above the resulting diagram is an implementation of the staggered quantum walk in the MBQC paradigm. As mentioned before, the specification of these gates allows for the reuse of qubits and, therefore, keeps the cluster state relatively compact.

The whole process can be generalized, much like as in [3], to a n-qubit protocol, starting from the n-qubit implementation of the alternative evolution operator depicted in Figure 6.

The configurations obtained for both three qubits and an arbitrary number of qubits, are depicted below in Fig. 13 and Fig. 14, respectively. This concludes the derivation of implementations of staggered quantum walks for the MBQC paradigm.

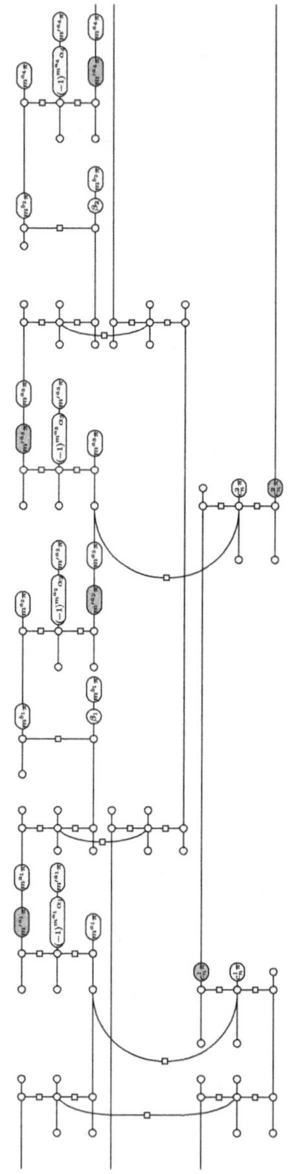

Figure 13: Reconfiguration of the 3 qubit case (Fig.5) as a MBQC protocol.

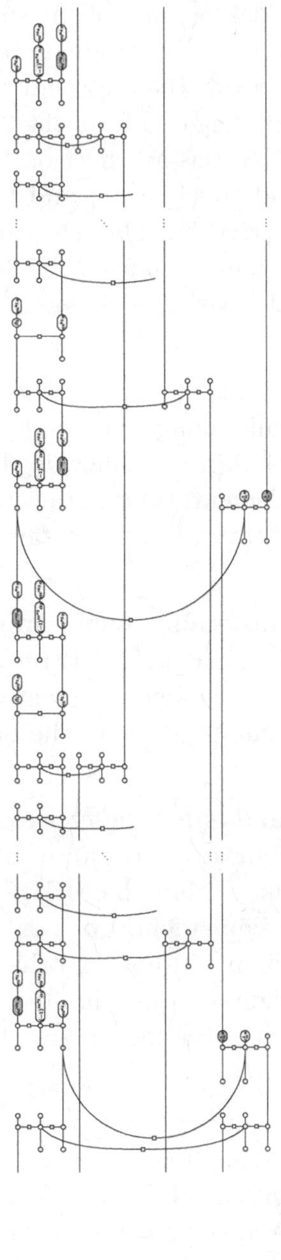

Figure 14: Reconfiguration of the general case (Fig.6) as a MBQC protocol.

However, a few notes are in order.

First note that the size of the cluster state must be $3n$, where n is the number of qubits in the original circuit model for the specific staggered quantum walk at hands. This is because each single qubit in the original circuit model is represented by 3 nodes in the cluster state. Moreover, the quantum state of a given qubit in the nth position (indexes start at 0) is propagated from the $3n^{th}$ node to the $(3n+2)^{th}$ node (or vice-versa) with every gate that has an effect on the nth qubit. As a consequence, every qubit between the second (position 1) and the $(n-2)^{th}$, has the quantum state in the same node as it started in. The first qubit (position 0) and last qubit, $(n-1)^{th}$, always swap the quantum state from the $3n^{th}$ node to the $(3n+2)^{th}$ node (or vice-versa). This is due to the parity of the number of operators applied to these qubits.

Another relevant remark is that, any set of three nodes that represent the same qubit can be mirrored horizontally along the middle one. This allows for the correct composition of these operators and in turn allow for longer staggered quantum walks. As a consequence, in this implementation any even length prefix of the quantum walk will have the quantum states in the exact same positions as they were at the beginning of the quantum walk.

Although the process of transforming quantum gate protocols into MBQC protocols is not novel ([12]), the derivations (11, 12) utilizing the techniques described above and the implementation of a staggered quantum walk within the paradigm of measurement based quantum computing is, to the best knowledge of the authors, a novel endeavour.

Even though this process allows for the conversion of quantum protocols, the results for this specific reconfiguration are quite difficult to compare to other implementations of quantum walks within the MBQC paradigm. To the best of our knowledge there are not other implementations of the staggered quantum walk in the measurement based paradigm. Moreover other implementations of quantum walks within the MBQC paradigm are quite limited, and most ([9],[14]) of them are fundamentally different in terms of the mechanics of the quantum walk.

If we were to compare the MBQC implementation to the gate model implementation based on the resources necessary to compute this algorithm it would be strictly worse, due to the necessity of having 3 nodes in the graph-state to represent a single qubit. However, when dealing with a MBQC protocol there are some extra assurances that can be taken into consideration, such as the increased resilience to quantum decoherence. In particular are very energy efficient compared to other quantum architectures such as superconducting/trapped-ion quantum computers.

6 Conclusions and Future Work

This paper discussed intra and inter-paradigm reconfiguration of quantum programs, formalised as *graph reconfiguration* processes. The process was illustrated through a non trivial example: staggered quantum walks. In a first moment the original circuit was reconfigured within the circuit based model for quantum computation, leading to a much more compact, alternative evolution operator.

This procedure was then extended so that, through a systematic, but cautious application of ZX rules, the optimized circuit lead to an implementation in a completely different paradigm, that of measurement-based quantum computation.

Reasoning within the measurement based paradigm is not entirely intuitive. Therefore, the approach proposed in this paper, in which an implementation is obtained through a reconfiguration process, seems a good alternative to start from a direct implementation attempt in the measurement based paradigm. Indeed, and similarly to what is done in [12], our approach starts from a circuit-based model, which is first optimised (*intra*-paradigm reconfiguration) and them reconfigured towards an implementation in the measurement based paradigm (*inter*-paradigm reconfiguration).

As current/future work we are extending this exercise to other well-known quantum protocols and algorithms. With possibly a broader impact, we are developing suitable dynamic logics over ZX models and reconfigurations to be able to state and verify relevant properties along the reconfiguration process.

References

[1] Bob Coecke and Ross Duncan. Interacting quantum observables: categorical algebra and diagrammatics. *New Journal of Physics*, 13(4):043016, April 2011.

[2] Niel de Beaudrap, Aleks Kissinger, and John van de Wetering. Circuit extraction for ZX-diagrams can be #P-hard. In Mikolaj Bojanczyk, Emanuela Merelli, and David P. Woodruff, editors, *49th International Colloquium on Automata, Languages, and Programming, ICALP 2022, July 4-8, 2022, Paris, France*, volume 229 of *LIPIcs*, pages 119:1–119:19. Schloss Dagstuhl - Leibniz-Zentrum für Informatik, 2022.

[3] Bruno Jardim, Jaime Santos, and Luís S. Barbosa. Reconfiguring staggered quantum walks with zx. In José Proença, Raul Fervari, Manuel A. Martins, Reinhard Kahle, and Graham Pluck, editors, *Software Engineering and Formal Methods. SEFM 2024 Collocated Workshops*, pages 112–119, Cham, 2025. Springer Nature Switzerland.

[4] Aleks Kissinger and John van de Wetering. PyZX: Large scale automated diagrammatic reasoning. *Electronic Proceedings in Theoretical Computer Science*, 318:229–241, May 2020.

[5] Aleks Kissinger and John van de Wetering. Reducing the number of non-clifford gates in quantum circuits. *Physical Review A*, 102(2), August 2020.

[6] Aleks Kissinger and John van de Wetering. *Picturing Quantum Software: An Introduction to the ZX-Calculus and Quantum Compilation.* Preprint, 2024.

[7] R. Portugal, M. C. de Oliveira, and J. K. Moqadam. Staggered quantum walks with Hamiltonians. *Physical Review A*, 95(1), January 2017.

[8] Renato Portugal, Raqueline A. M. Santos, Tharso D. Fernandes, and Demerson N. Gonçalves. The staggered quantum walk model. *Quantum Inf. Process.*, 15(1):85–101, 2016.

[9] Xiaogang Qiang, Xiaoqi Zhou, Jianwei Wang, Callum M. Wilkes, Thomas Loke, Sean O'Gara, Laurent Kling, Graham D. Marshall, Raffaele Santagati, Timothy C. Ralph, Jingbo B. Wang, Jeremy L. O'Brien, Mark G. Thompson, and Jonathan C. F. Matthews. Large-scale silicon quantum photonics implementing arbitrary two-qubit processing. *Nature Photonics*, 12(9):534–539, August 2018.

[10] Robert Raussendorf, Daniel Browne, and Hans Briegel. The one-way quantum computer–a non-network model of quantum computation. *Journal of Modern Optics*, 49(8):1299–1306, July 2002.

[11] Jaime Santos. Quantum random walks: simulations and physical realizations. MSc Thesis in Engineering Physics, DI, Universidade do Minho, 2021.

[12] Tobias Stollenwerk and Stuart Hadfield. Measurement-based quantum approximate optimization. In *IEEE International Parallel and Distributed Processing Symposium, IPDPS 2024 - Workshop, San Francisco, CA, USA, May 27-31, 2024*, pages 1115–1127. IEEE, 2024.

[13] Salvador Venegas-Andraca. Quantum walks: a comprehensive review. *Quantum Information Processing*, 11(5):1015–1106, July 2012.

[14] Tianyi Wang, Xiaoguang Chen, and Jianxiong Liang. Four-qubit cluster states generation

through multi-coin quantum walk. *Applied Sciences*, 12(17), 2022.

Received July 2025

www.ingramcontent.com/pod-product-compliance
Lightning Source LLC
LaVergne TN
LVHW060451061126
851857LV00016B/1788